CONTENTS

USING THIS BOOK

This Book Was Designed for You —

a talented, yet busy teacher. We know that you want to provide students with the most interesting and comprehensive units of study possible. We also know how much time it takes to fully prepare to teach a topic. That's why we developed the **Ten Easy Steps** series. We've covered all the bases. From planning to implementation — it's all here.

STEP 1 Using This Book
This section contains a little background information on how to use this book and a peek into what you'll be teaching during the 10 lessons.

STEP 2 Gather Great Resources
In this section, you'll find a list of books to use when teaching yourself and others about the Solar System, a list of web sites that help explain the space topics you'll be teaching, and a list of field trip and guest speaker ideas. There's even a letter for parents to help you find a great speaker!

STEP 3 Speak the Lingo
This is where you'll find all the vocabulary words and definitions specific to the topics covered in this book as well as worksheets and pocket chart ideas designed to reinforce the vocabulary.

STEP 4 Set the Scene
It's important to set the tone for the unit of study. This means transforming your classroom environment to reflect the concepts being taught. In this section, you'll find great ideas for interactive learning areas and classroom decoration.

STEP 5 Plan a Project
In this section, you'll find plans for an ongoing project students will be working on throughout the unit of study. It's a great way to apply what they're learning each day.

STEP 6 Teach Ten Terrific Lessons
Ten complete lessons can be found within this section. Each lesson includes essential concept information, experiments, hands-on learning activities, journal prompts, homework ideas, and teaching notes on each experiment.

STEP 7 Cross the Curriculum
Take one great concept, teach it in multiple curriculum areas, and you're sure to reinforce learning. In this section, you'll find ways to extend the learning across all areas of the curriculum, including social studies, reading, writing, math, and art.

STEP 8 Tie in Technology
In this section, we provide you with ideas and project planning pages for a multimedia presentation and web site creation.

STEP 9 Assess Learning
This section provides a variety of assessment options. But don't wait until the end of the unit to assess your students. This book is filled with journal and homework ideas to assess from the start.

STEP 10 Celebrate!
Once you've completed a unit of study as compelling as this, you'll want to celebrate. In this section we've provided an idea for a great end-of-unit celebration.

A Note About the Internet
The Internet is a constantly changing environment. The sites listed as additional references were current at the time this book went to press.

INTRODUCTION TO THE SOLAR SYSTEM

Learning about the Solar System involves so much more than simply memorizing the names of the planets in order, knowing that Mars is called the Red Planet or that Saturn has rings. In this unit of study, students will learn all the parts that make up our great Solar System and our missions to space!

LESSONS

Each of the following lessons in **Step 6** features a quick informative mini-lesson, fascinating facts, easy-to-accomplish experiments and activities, a journal prompt, and a homework idea.

1. **What is the Solar System?**
   Objective: To learn about the elements that make up our Solar System and how these elements interact.

2. **The Sun**
   Objective: To investigate properties of the Sun and understand its role in our Solar System.

3. **Tilted, Rotating & Revolving!**
   Objective: To understand seasons, day and night, and our calendar.

4. **Inner Planets: Mercury, Venus, Earth & Mars**
   Objective: To investigate properties and characteristics of Mercury, Venus, Earth, and Mars.

5. **Outer Planets Part 1: Jupiter & Saturn**
   Objective: To investigate properties and characteristics of Jupiter and Saturn.

6. **Outer Planets Part 2: Uranus, Neptune & Pluto**
   Objective: To investigate properties and characteristics of Uranus, Neptune, and Pluto.

7. **The Moon**
   Objective: To learn about Earth's Moon, its characteristics, its phases, and its effect on Earth.

8. **Our Night Sky**
   Objective: To learn about stars, constellations, comets, and galaxies.

9. **History of Astronomy**
   Objective: To learn about early astronomers and their methods, theories, and discoveries.

10. **Space Exploration**
    Objective: To learn about the National Aeronautics and Space Administration (NASA) and the history of space travel.

OTHER TOOLS

In addition to lessons and experiments, this book contains many other tools to help you make this unit more complete, including:

- A list of books and web sites for you and your students. **(Step 2)**
- A vocabulary list of Solar System words and definitions, along with vocabulary worksheets, puzzles, and pocket chart activities. The back of the book contains a pocket chart card for each vocabulary word. You can use the pocket on the inside back cover to store the cards once they're torn out from the book. **(Step 3)**
- Learning center ideas filled with information to help you set up a classroom Solar System reporting center. **(Step 4)**
- An ongoing project in which students create a scale model of the Solar System complete with a tape recording of information. **(Step 5)**
- Cross-curricular learning ideas to carry the study of the Solar System into other areas of your curriculum. **(Step 7)**
- Connections to technology via a Solar System FAQ's web page project and a multimedia presentation on the planets. **(Step 8)**
- Assessment tools including rubrics, journals, and tests. You'll find plenty of tools and ideas for alternative or traditional assessment of student learning. **(Step 9)**
- A celebratory end-of-the-unit event that allows students to "show what they know" while reinforcing the content covered. **(Step 10)**

Great Resources for You

It's impossible to be an expert on every subject you teach, yet that's exactly how your students see you. Before you begin teaching this Solar System unit, spend a few nights reviewing the following web sites and books, and you'll be up to speed in no time!

Web Sites

Windows to the Universe
http://msgc.engin.umich.edu/

This incredible site contains an entire curriculum on the universe! The stated purpose of the site is to "develop a fun and different web site about the Earth and space sciences." It accomplishes that goal and much more!

The Nine Planets
http://www.nineplanets.org/

This is one of the best Solar System sites on the Internet! Each page has text and images relating to the planet and its moons (if any), and some have sounds and movies. Most pages provide links to additional related information.

National Science Teachers Association
http://www.nsta.org/

This organization's site has super resources to give your ideas a boost. It includes a wonderful listing of science and math links.

Learning Resources
http://www.learningresources.com

Seek out this site for a list of **10 Steps**-recommended web sites or great products for your classroom. You'll want to head to *Activities & Resources* for the list.

Books That Help Prepare

Asimov, Isaac. *The Earth's Moon.* *Milwaukee, Wisconsin: Gareth Stevens Inc, 1988.*

Written by a true master of the subject, this book provides all the information you'll need to be an "expert" on Earth's Moon.

Lippincott, Kristen. *DK Eyewitness: Astronomy.* *New York: DK Publishing, 1999.*

Like all *Eyewitness* titles, this one is full of great photos and concise information divided into easy-to-understand subtopics. This book will easily get you up to speed and provide great bits of interesting information not always included with a study of the Solar System.

Mitton, Simmon. *The Young Oxford Book of Astronomy.* *New York: Oxford University Press, 1995.*

While the text in this book may be too in-depth for student reading, the information will give you a nice overview before beginning to teach a unit on space. In addition, the graphics could be useful as you explain information to students.

Schatellow-Sawyer, Bonnie. *Exploring the Planets.* *New York: Scholastic Professional Books, 2000.*

This book features a discovery-based learning style — posing questions, then providing information and activities — to help students find answers. The book also includes a great poster for you to hang on the wall!

Great Resources for Your Students

Surrounding your students with great resources is a sure way to stimulate learning. The first step is to encourage your students to take a look at a few of the great web sites and books listed on this page and on page 7. The field trip ideas in this section will also get your students in gear for a study of the Solar System. You'll have a captive audience before you even begin teaching!

Web Sites

The Solar System

http://www.jpl.nasa.gov/solar_system/solar_system_index.html

JPL (Jet Propulsion Laboratory) and NASA have put together this straightforward site with information on the planets. Each link has descriptive information about the planet along with information about interplanetary exploration.

An Overview of the Solar System

http://www.seds.org/billa/tnp/overview.html

This is a great place to start your exploration! It has everything from simple diagrams of the inner and outer planets to planetary classification systems and detailed photographs. Dozens of links provide access to further information.

The Nine Planets

http://www.nineplanets.org/

This is one of the best Solar System sites on the Internet! Each page has text and images relating to the planet and its moons (if any), and some have sounds and movies. Most pages provide links for additional related information.

Views of the Solar System

http://www.solarviews.com/eng/homepage.htm

This site provides a vivid multimedia adventure as it takes the viewer on a virtual journey to the Sun, planets, moons, comets, asteroids, and more. Discover the latest scientific information, or study the history of space exploration, rocketry, early astronauts, space missions, and spacecrafts through a vast archive of photographs, scientific facts, text, graphics, and videos.

Our Solar System: The Planets and Their Motion

http://www.athena.ivv.nasa.gov/curric/space/planets/index.html

Take a tour of the planets with the pictures and online movies on this site. Try playing the movies at a slow speed, and notice the differences in the time it takes for each planet to revolve around the Sun (its year). Have your students draw pictures and match each planet with its name.

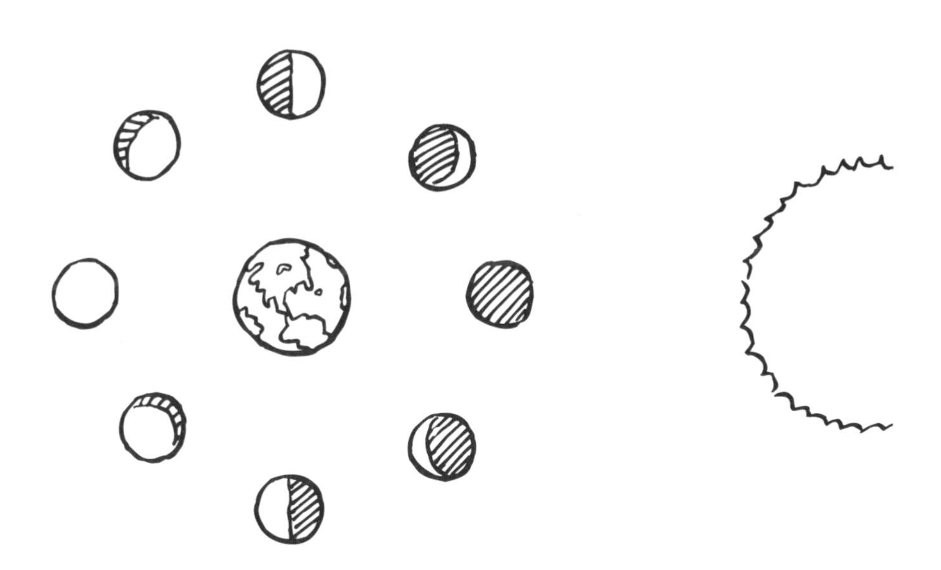

Great Resources for Your Students

Books

Brenner, Barbara. ***Planetarium: The Museum that Explores the Many Wonders of Our Solar System.*** *New York: Bantam, 1993.*

The format of this book makes it a very "kid-friendly" resource. The information gives students an inside look at the Solar System by taking the reader on an actual walk through space. The book also includes a number of easy-to-complete experiments.

Cole, Joanna. ***The Magic School Bus — Lost in the Solar System.*** *New York: Scholastic, 1990.*

Ms. Frizzle is up to her old tricks! This time her field trip takes students on a journey through our Solar System. As with all the other "Magic School Bus" titles, this delivers the facts with a generous dose of humor.

Kerrod, Robin. ***The Solar System.*** *Minneapolis: Lerner Publications, 2000.*

This book covers all aspects of the Solar System, but does a particularly nice job of explaining the planets, how they rotate, and their place in the Solar System.

Lauber, Patricia. ***Journey to the Planets.*** *New York: Crown Publishers, 1993.*

The photographs in this book are amazing! There's not a single hand-drawn image. It's a great book to use during instruction.

Schatellow-Sawyer, Bonnie. ***Exploring the Planets.*** *New York: Scholastic Professional Books, 2000.*

This book features a discovery-based learning style — posing questions, then providing information and activities — to help students find answers. The book also includes a great poster for you to hang on the wall!

Scott, Elaine. ***Adventure in Space.*** *New York: Hyperion, 1995.*

This book gives a play by play of the mission to repair the Hubble Telescope. It starts with an explanation of the problems with the telescope and goes on to show the crew, their training, and the ultimate success of the mission.

Simon, Seymour. ***Our Solar System.*** *New York: Morrow Junior Books, 1992.*

This book does a wonderful job covering each of the planets as well as includes general information about our Solar System. The bonus to this book is the beautiful, up close and personal photographs of the planets.

Stott, Carole. ***Night Sky.*** *New York: DK Publishing, 1993.*

Your students may just seek out this book on their own. Its small size and limited, yet informative text, make it an easy-to-read favorite. It provides a straightforward explanation for all the Solar System concepts you'll be covering in this unit of study.

Guest Speaker Ideas

1. A local science center or planetarium employee.
2. An astronomy professor from a local college or university.
3. A parent or teacher who is an avid telescope user and night sky enthusiast.

Field Trip Ideas

1. Create your own online field trip using a few of the sites listed in this text. Require students to fill out field trip information sheets or gather specific bits of information as they visit each site.
2. Host a night sky exploration on your school's front lawn.
3. Visit a local science center or planetarium.

Letter to Parents

Dear Parents,

Over the next few weeks our class will be studying the Solar System. Our topics of interest will include:

1. **What is the Solar System?**
2. **The Sun**
3. **Tilted, Rotating & Revolving**
4. **Inner Planets: Mercury, Venus, Earth & Mars**
5. **Outer Planets Part 1: Jupiter & Saturn**
6. **Outer Planets Part 2: Uranus, Neptune & Pluto**
7. **The Moon**
8. **Our Night Sky**
9. **History of Astronomy**
10. **Space Exploration**

If you have personal stories or insights to share on any of the above listed topics, we would love to have you come in and talk to the class. We would also appreciate any materials (books, videos, and posters) that you'd be willing to share for the next few weeks.

Reinforcing learning at home will help your child retain the information learned in school. Try to find time to discuss the topics, ask questions, and stay involved with homework and projects. If possible, explore the following web sites with your child.

Windows to the Universe
http://www.windows.ucar.edu/

The Solar System
http://www.jpl.nasa.gov/solar_system/solar_system_index.html

The Nine Planets
http://www.nineplanets.org/

Thank you for all your help and support.

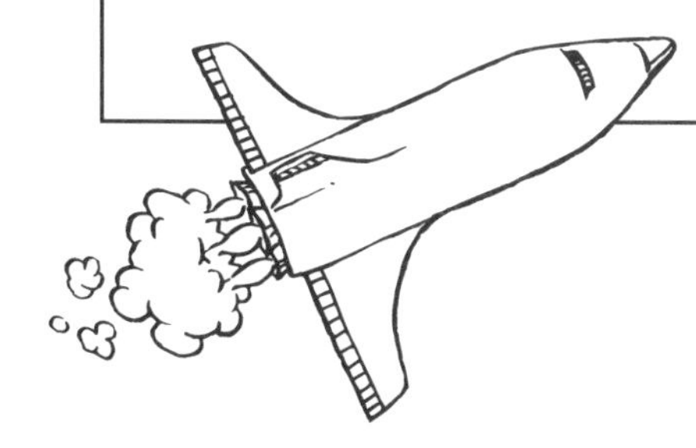

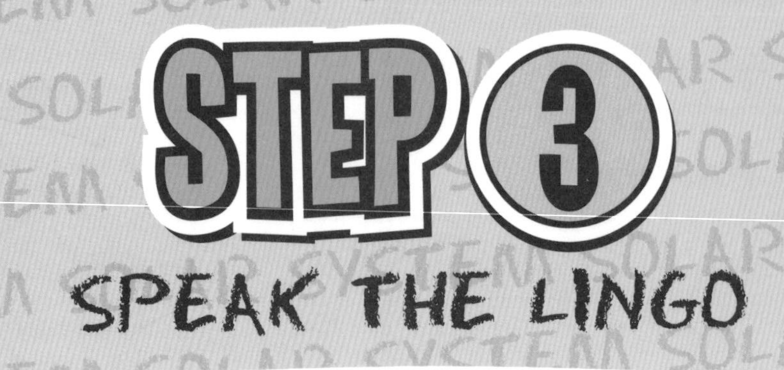

Solar System Vocabulary

Understanding the meanings of key words before delving into the topic will help students grasp the concepts later on. The pages in **Step 3** provide the practice to help students retain the words and their definitions. The worksheets are based on the following list of vocabulary words from the lessons in **Step 6.** Each word is also printed on the pocket chart cards located at the end of this book.

Lesson 1

Asteroid Belt
concentrated group of rocks that revolves around the Sun between the inner planets and outer planets

inner planets
planets closest to the Sun: Mercury, Venus, Earth, and Mars

outer planets
planets farthest from the Sun: Jupiter, Saturn, Uranus, Neptune, and Pluto

Solar System
grouping in space that revolves around the Sun and includes planets, asteroids, meteors, and comets

orbit
path a planet follows as it moves around the Sun

revolve
to circle around the Sun (one revolution around the Sun equals one year)

Lesson 2

Sun
large star that is the center of our Solar System

sunspots
storms on the Sun's surface

Lesson 3

axis
imaginary line that runs through the center of the Earth, and around which the Earth rotates on a slight tilt

rotate
to spin or turn on an axis (Earth rotates once every 24 hours)

seasons
variations in temperature and daylight conditions resulting from Earth's tilt on its axis and its orbit around the Sun

sundial
early instrument used to measure time

year
time it takes for Earth to complete one revolution around the Sun

Lesson 4

Earth
third planet from the Sun, able to sustain life

Mercury
planet closest to the Sun

Venus
second planet from the Sun, surrounded by a deadly carbon dioxide atmosphere

Mars
fourth planet from the Sun and closest to Earth, has a rough crater-like surface

Lesson 5

Jupiter
fifth and largest planet from the Sun, has a giant red spot

Saturn
sixth planet from the Sun, has numerous rings

Lesson 6

Neptune
eighth planet from the Sun, covered by cloud layers

Pluto
ninth planet from the Sun, is the smallest planet

Uranus
seventh planet from the Sun, called the sideways planet

Lesson 7

craters
indentations on the surface of the Moon

Moon
satellite of Earth

Moon phases
changes in the Moon's appearance over a month caused by changing amounts of reflected sunlight

lunar eclipse
occurs when the Moon passes through Earth's shadow

solar eclipse
occurs when the Moon passes between the Sun and Earth

Lesson 8

asteroids
rocky bodies smaller than planets that orbit the Sun, often called minor planets

astronomer
person who studies the sky and its objects

Big Dipper
highly visible part of a larger constellation, often used to locate the North Star

black hole
space from when a star dies

comet
large, lightweight body that orbits the Sun; contains a head and a tail

constellation
grouping of stars that resembles a recognizable image or shape

galaxy
large grouping of stars

luminous
able to give off light, such as stars

meteoroid
small pieces that break off larger bodies like planets and asteroids

meteor
streak of light that occurs when a meteoroid enters Earth's atmosphere

meteorite
meteoroid that reaches the Earth without burning up in the atmosphere

planetarium
building designed for the study of stars and planets

star
hot, glowing ball of gas that gives off its own light

Lesson 9

Aristotle
thought Earth was the center of the Solar System

Copernicus
realized the movement of the stars was due to Earth's motion; believed Earth moved around the Sun

Galileo
built a telescope for observing the night sky

Hypatia
one of the first women astronomers

Newton, Sir Isaac
famous for his studies on gravity

Ptolemy
realized Earth was round, claimed all the other planets move around Earth

Lesson 10

Armstrong, Neil
first man to walk on the Moon

Friendship 7
first U.S. spacecraft to orbit Earth

Gargarin, Yuri
first person to circle Earth in a spacecraft

Glenn, John
first American to circle Earth in a spacecraft

Space Shuttle
first spaceship designed to be used more than once

Space Station
international orbiting laboratory for studying our Solar System

Sputnik 1
first man-made satellite to be placed in orbit (Russian)

Pocket Chart Vocabulary Activities

Using your pocket chart cards and a pocket chart, try a few of the activities listed below to introduce and develop Solar System vocabulary words.

Begin Each Lesson

Begin each lesson by showing the new vocabulary words that apply. At the end of each lesson, review the words with your students.

What's the Word?

Divide the class into teams. Pull one vocabulary card, and give its definition without showing the face of the card. The first team to "buzz in" with the correct word receives a point. Continue until all the cards have been revealed.

Name that Planet

Use the planet picture cards as flash cards. Challenge students to be first to identify the planets as you flash their pictures.

Definition, Please

Play "What's the Definition, Please?" Place all the cards facedown in the pocket chart. Divide the students into four teams. Teams take turns sending a player up to the chart to retrieve a card to take back to his or her group. The group then has 30 seconds to come up with a definition for the word to receive a point. If the group cannot come up with the definition, the other teams have the opportunity to answer. The first team to "buzz in" with the correct definition gets the point, and regular play resumes with the next team going up to draw a card. Continue until all the terms have been defined.

Name ______________________________

Solar System Vocabulary Practice

Fill in the blanks with the correct word from your Solar System vocabulary word sheet or pocket chart words.

1. The seasons occur as a result of Earth's tilt on its ______________________.

2. ______________________ is usually the furthest planet from the Sun.

3. ______________________ is often called the "Red Planet."

4. Rings are the most famous feature of the planet ______________________.

5. ______________________ created a telescope for watching the night sky and made numerous discoveries about Jupiter.

6. A ______________________ is a grouping of stars that forms an image.

7. The ______________________ is the center of our universe.

8.  ______________________ are small and large rocks that orbit the Sun.

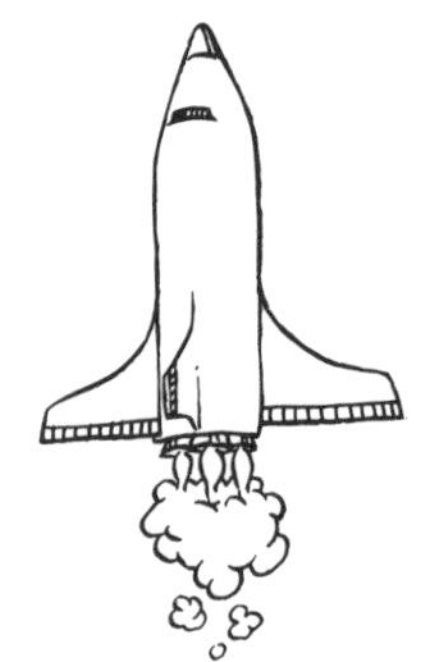

9. Jupiter, Saturn, Neptune, Uranus, and Pluto are referred to as the ______________________ planets.

10. Mercury, Venus, Earth, and Mars are referred to as the ______________________ planets.

11. Both the Moon and Mars have numerous ______________________, or indentations, on their surface.

Name ______________________________

Solar System Crossword Puzzle

Fill in the crossword puzzle using these clues.

Across

1. to circle around the Sun
2. large and small pieces of rock floating in space
3. nine major bodies found in our Solar System
4. Italian astronomer who created a telescope for looking at the night sky
5. place to go to observe the stars

Down

6. called the Red Planet
7. to spin or turn on an axis
8. eighth planet from the Sun
9. planet known for its beautiful rings
10. the largest planet

SET THE SCENE

Classroom Learning Centers

Just as backdrops and costumes are important to a play, a welcoming classroom environment is important to foster learning. The room should be fun, inviting, and interactive. With that in mind, this section features learning center activities and bulletin board ideas to help you set up the room for a unit on the Solar System.

1. Constellation Creation Learning Center

Provide copies of zodiac pictures that are also constellations (e.g., twins, crab, scorpion, ram, bull, fish, etc.). Upon visiting this center, students should choose the constellation that represents their zodiac sign and draw its stars on a piece of cardboard. Students can then use a pushpin to carefully poke holes in the cardboard where each star is placed. Provide a dark place and a flashlight for students to shine behind the cardboard, making the stars "shine."

2. Research Timeline Learning Center

In this center, history and science work hand in hand. Provide students with multiple books containing information on the history of space exploration and current missions.

Place long sheets of bulletin board paper as well at the center. Then, have students work in pairs to create a timeline of events that have played a significant role in our understanding of the Solar System.

3. Sun Facts Learning Center

The Sun is the center of our Solar System, so it deserves a little extra attention. At this center, students will conduct research for a Sun fact, create a fact card, and add the card to the Sun bulletin board mentioned on page 16. Be sure to keep books on the Sun at this center as well as index cards.

4. Experiment Learning Center

This center will help you organize all the experiments in this book for your students. Be sure to have the following materials at this center:

- supplies for the experiments in **Step 6**
- instructions for completing experiments
- experiment **Science Logs**

Encourage students to record their results in their journals or on the Science Logs for each experiment. You may also want to include directions for other experiments you've come across during your research.

Learning Centers Checklist: Teachers

Use the narrow column to the left of the activity title to record the date the student completed the activity. In the activity columns, record a grade or symbol to reflect the level of completion. You might also use the activity columns to jot notes about the student's performance.

Student		1. Constellation Creation		2. Research Timeline		3. Sun Facts		4. Solar System Experiments

Learning Centers Checklist: Students

Photocopy this page for each student and cut it in half. Have your students use this sheet to get sign-off by you or a peer each time they successfully complete a center. Remind students that completing more than one center a day or repeating a center during the week is permitted.

Name ________________________________ Date ______________

Centers Week ____ – ____	Monday	Tuesday	Wednesday	Thursday	Friday
1. Constellation Creation					
2. Research Timeline					
3. Sun Facts					
4. Solar System Experiments					

Name ________________________________ Date ______________

Centers Week ____ – ____	Monday	Tuesday	Wednesday	Thursday	Friday
1. Constellation Creation					
2. Research Timeline					
3. Sun Facts					
4. Solar System Experiments					

Classroom Bulletin Board

The bulletin board ideas will help you and your students set up the room. Aside from these bulletin board ideas, you will find that posters of the planets, diagrams of the Solar System, and space flight images will be quite useful, especially when answering questions that come up during the lessons.

What's Out There? Bulletin Board

This bulletin board will serve as a reference point for the unit of study. Students can help add to it as they complete each lesson about the Solar System. Each time you study an element, add it to the board. The final board should include the Sun, each of the planets, stars, asteroids, galaxies, artificial satellites, and rockets. As students add images to the bulletin board, have them include a small card with information about the image.

Our Favorite Star! Bulletin Board

Use this bulletin board as an opportunity to highlight the importance of the Sun. Students will be conducting research for "fun Sun facts" at the **Sun Facts Learning Center**. Encourage them to use this board to include these facts about stars and how our Sun compares to other stars. Allow students to create artwork or bring in various images of the Sun to include on the board as well.

Scale Model of the Solar System

Requiring students to put their knowledge and skills to work is a great way to ensure long-term retention of content. In **Step 5: Plan a Project**, students have an opportunity to observe and gather information over a long period of time and share their data. The end product is an interactive scale model of the Solar System. Students also create an accompanying tape that gives listeners information about the model.

This project can be done individually, but it would also work well if student pairs coordinated research, recorded data, and created the model. Be sure to follow these steps in order with your students.

1. Create Planet Models

Decide if the students will be working individually on this project or in pairs. Also choose the materials to model each planet. If you would like to have students represent a 2-dimensional model, have students use construction paper. A 3-dimensional model can be quite fun and interesting for students, too. In this case, you would want to use colored clay. You'll also need to provide these materials for each student or student pair:

- compass
- adhesive labels
- wire
- Styrofoam™ or string
- ruler or measuring tape (optional)

When creating a scale model, remind students that the planets should be relative in size to one another. But, they certainly wouldn't want to build Jupiter to size! Therefore, we've included a few guidelines for your students to use when building their scale model of the planets and the Sun. The dimensions of each planet should follow the proportions shown in the table to the right. It is in inches. You'll want to provide large sheets of construction paper for the larger planets like Jupiter, Saturn, Uranus, and Neptune.

Note: Remind students that when using a compass to create the circles, they must set the compass at half the diameter, called its radius. They may want to check their work with a ruler.

After students are done making their models, have them label the planets using adhesive labels.

Planet	Diameter	Radius
Mercury	1″	½″
Venus	3″	1½″
Earth	3″	1½″
Mars	1½″	¾″
Jupiter	33½″	16¾″
Saturn	28½″	14¼″
Uranus	12″	6″
Neptune	11½″	5¾″
Pluto	½″	¼″

Scale Model of the Solar System (continued)

In this second part of the project, students will have the opportunity to place their planets accordingly in the Solar System and present their findings to the class. Encourage students to come back to this project as you are covering a part of it during the lesson.

2. Place Each Planet

Have students use a Styrofoam™ base and wire to place the planets in order. You may also choose to use string to hang the planets from the ceiling.

If you have the space, and would like to place the planets in order, the scale below will help you. According to this scale, 1 mm equals 1 million km.

Planet	Distance from Sun
Mercury	58 mm
Venus	108 mm
Earth	150 mm
Mars	228 mm
Jupiter	778 mm
Saturn	1,430 mm
Uranus	2,870 mm
Neptune	4,500 mm
Pluto	5,900 mm

3. Do Research

Now that students have created a Solar System, encourage them to learn more about each planet and the Sun. Make 10 copies of the **Planet Information Sheet** on page 20 for each student or pair. Allow plenty of time for students to research and record information.

3. Write a Script and Rehearse

Have your students use their **Planet Information Sheets** to write a short script containing information about each planet, the Sun, and our Solar System. Remind students that the script will be taped and will be played as others view their Solar System model. The tape should be lively and entertaining as well as educational. Allow time for students to rehearse reading their scripts.

4. Record Scripts

Help students use a tape player to record the Solar System information in their scripts. Remind students to listen to their final product and make corrections, if necessary, before turning it in for a grade.

5. Share With Others!

Set aside a day for viewing the final products and listening to the tapes. Encourage peer review and assessment.

Name ____________________

Weekly Project Goals

Use this checklist to help you make plans for your model, spend time researching planet information, write a script, and tape record your information. You can record the work done as you go along. Ask your teacher for additional copies of this checklist if necessary.

Project Tasks & Goals for the week of ____ – ____	Monday	Tuesday	Wednesday	Thursday	Friday
Building Model Planets					
Building the Model Scale					
Researching Planets					
Writing a Script					
Tape Recording the Information					
Other:____________________					

Planet Information Sheet

Complete one sheet for each planet and an additional sheet for the Sun. Use these pages as you prepare the script for your interactive scale model of the Solar System.

Planet Name: ______________________

Planet Position: ______________________

Distance from the Sun: ______________________

Diameter: ______________________

Length of One Year: ______________________

Length of One Day: ______________________

Satellites (Moons): ______________________

Number of Satellites: ______________________

Name(s) of Satellites: ______________________

Interesting Facts:

____

____

____

____

____

____

____

____

TEACH TEN TERRIFIC LESSONS

Introduction

The 10 lessons presented on the pages that follow provide a comprehensive study of the Solar System. Work through the steps in order or pick and choose the activities that will enhance what you're already teaching — the choice is yours!

Each lesson contains 3 parts:

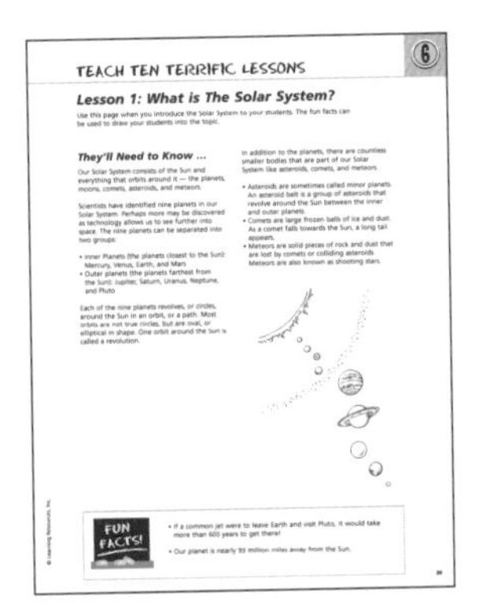

1. Teacher Note Page(s)
Provides a general overview of the lesson's topic. These pages include:
- **They'll Need to Know** ... for a general overview of the lesson's topic
- **Prove It!** for points to bring up as students are working through the experiments
- **Journal Prompt** to assess student learning and to give students the opportunity to put the science concept into their own words and/or expand their thinking beyond the topic
- **Homework Idea** to follow up on the concept at home

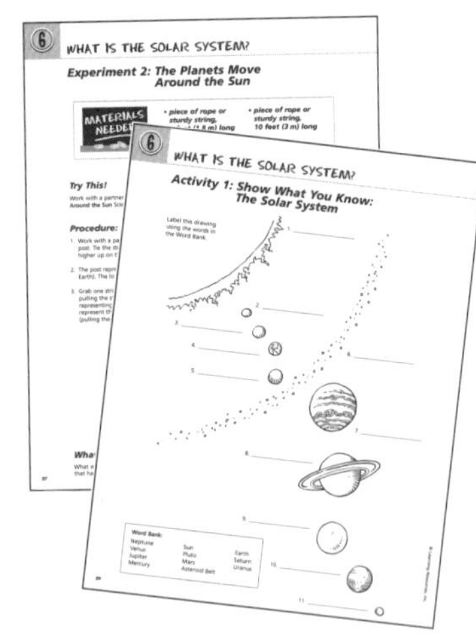

2. Experiments and Activities
Provides hands-on experiences designed to reinforce the day's lesson. The teaching notes page provides background information for each experiment.

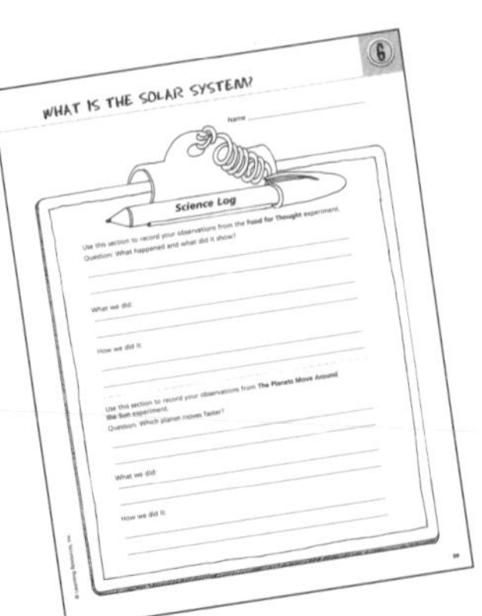

3. Science Log
Provides a space for students to record the concepts learned and their observations from the experiments.

Overview

The following table explains the objective of each lesson as well as the experiments, activities, and supplies needed in each lesson. Be sure to collect these supplies in advance.

Lesson	Supplies
1. What is the Solar System? Students learn about the elements that make up our Solar System and how these elements interact.	**Experiment 1: Food for Thought:** grapefruits, oranges, plums, peas, peppercorns **Experiment 2: The Planets Move Around the Sun:** rope or sturdy string, sturdy post or pole **Activity 1: Show What You Know: The Solar System:** page 29
2. The Sun Students investigate properties of the Sun and understand its important role in our Solar System.	**Experiment 1: Safe "Sun" Glasses:** cardboard, white paper, pen, scissors, tape, binoculars, stool or ladder **Experiment 2: A Plant Needs Sunshine:** identical potted plants, cardboard or construction paper, scissors, tape **Activity 1: Show What You Know: the Sun:** worksheet, page 34
3. Tilted, Rotating & Revolving Students work through activities on seasons, day and night, and our calendar.	**Experiment 1: Understanding Day, Night, and Seasons:** lamp or flashlight, globe, tape, small paper doll **Activity 1: Make a Sundial:** wood board, modeling clay, pencil, triangle or carpenter's square, magnetic compass, watch or clock **Activity 2: Show What You Know: Tilted, Rotating & Revolving:** page 39
4. Inner Planets: Mercury, Venus, Earth & Mars Students explore the properties and characteristics of Mercury, Venus, Earth, and Mars and compare the planets to other elements within the Solar System.	**Experiment 1: Create Mercury's Craters:** drop cloth or newspapers, large glass bowl, flour, loose dirt or sand, small stool or chair, marbles, rocks, small ball, small building blocks **Activity 1: Show What You Know: Planet Fact File:** page 44
5. Outer Planets Part 1: Jupiter & Saturn Students discover the characteristics of Jupiter and Saturn and compare the planets to other elements within the Solar System.	**Experiment 1: Create a Jupiter Storm:** large glass bowl, milk, measuring cup, eye dropper, yellow food coloring, red food coloring, dishwashing liquid **Activity 1: Show What You Know: Planet Fact File:** page 44

Overview *(continued)*

Lesson	Supplies
6. Outer Planets Part 2: Uranus, Neptune & Pluto Students learn the properties and characteristics of Uranus, Neptune, and Pluto and compare the planets to other elements within the Solar System.	**Experiment 1: Discover Uranus's Rings:** pencils, rectangular Styrofoam™ blocks, black paint, flashlights, rulers **Activity 1: Show What You Know: Planet Fact File:** page 44
7. The Moon Students research Earth's Moon, its characteristics, its phases, and its effect on our planet.	**Experiment 1: Understanding the Phases of the Moon:** flashlights, golf balls or Ping-Pong balls, tennis balls or softballs, aluminum foil **Experiment 2: Simulating a Solar Eclipse:** lamp, tennis balls **Activity 1: Show What You Know: The Moon:** page 57
8. Our Night Sky Students explore information on stars, constellations, comets, and galaxies.	**Experiment 1: Make a Constellation: Mini-Constellations** sheet (page 61), soup cans, black paper, pins, scissors, tape, flashlights, can opener **Activity 1: Show What You Know: Our Night Sky:** page 62
9. History of Astronomy Students learn about early astronomers and their methods, theories, and discoveries.	**Experiment 1: Newton's Third Law and Rockets:** drinking straws, scissors, oblong balloons (long and thin), string, fishing line, or sturdy thread, tape, butterfly clamps or paper clips **Activity 1: Astronomer Mini-Biography:** page 67, research materials
10. Space Exploration Students research NASA and the history of space travel.	**Experiment 1: Rockets in Space:** balloons, empty juice cans, measuring tape, drinking straws, string **Activity 1: Space Exploration Timeline:** scissors, glue, roll of paper, paint, markers, or crayons, research materials

Lesson 1: What is The Solar System?

Use this page when you introduce the Solar System to your students. The fun facts can be used to draw your students into the topic.

They'll Need to Know ...

Our Solar System consists of the Sun and everything that orbits around it — the planets, moons, comets, asteroids, and meteors.

Scientists have identified nine planets in our Solar System. Perhaps more may be discovered as technology allows us to see further into space. The nine planets can be separated into two groups:

- Inner Planets (the planets closest to the Sun): Mercury, Venus, Earth, and Mars
- Outer planets (the planets farthest from the Sun): Jupiter, Saturn, Uranus, Neptune, and Pluto

Each of the nine planets revolves, or circles, around the Sun in an orbit, or a path. Most orbits are not true circles, but are oval, or elliptical in shape. One orbit around the Sun is called a revolution.

In addition to the planets, there are countless smaller bodies that are part of our Solar System like asteroids, comets, and meteors.

- Asteroids are sometimes called minor planets. An Asteroid Belt is a group of asteroids that revolve around the Sun between the inner and outer planets.
- Comets are large frozen balls of ice and dust. As a comet falls towards the Sun, a long tail appears.
- Meteoroids are solid pieces of rock and dust that are lost by comets or colliding asteroids.

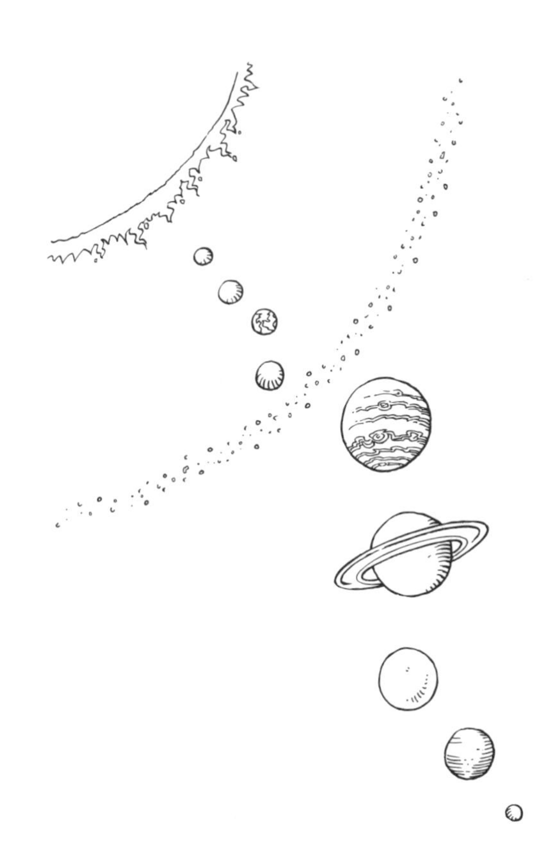

- If a common jet were to leave Earth and visit Pluto, it would take more than 600 years to get there!
- Our planet is nearly 93 million miles (150 million km) away from the Sun.

Lesson 1: What is The Solar System? (continued)

Everything in our Solar System revolves around the Sun. Compared to all the other bodies in our Solar System, the Sun is huge — more than 300,000 times larger than Earth. The Sun is the heaviest, largest, and hottest body in the Solar System. The entire Solar System is held together by the gravitational pull produced by the Sun.

Although the Sun is special to us, it's actually just another star in an even much larger system known as a galaxy. A galaxy is a large grouping of stars that spin and revolve around a central core. There are many different galaxies in our universe. Ours is called the Milky Way.

Prove It!

The **Food for Thought** experiment on page 26 uses food items to help students understand the scale and size of the various planets. Feel free to substitute similar items or provide non-food alternatives of the same size. You may also decide to have children work in pairs to keep food costs down.

The Planets Move Around the Sun experiment on page 27 works best if the person simulating an outer planet is a good bit taller than the person representing the inner planet.

With the activity on page 29, students will have an opportunity to label the celestial bodies in the Solar System.

Experiment 1: Food for Thought Teaching Notes:
It is often difficult for students to appreciate the actual size and volume of the planets. In this experiment, however, students were able to visualize the proportions of the planets within our Solar System. After students have completed this experiment, ask them, *Which planets make up the outer planets? Which planets make up the inner planets?* This will help reinforce the concepts taught.

Experiment 2: The Planets Move Around the Sun Teaching Notes:
With this experiment, students were able to clearly visualize the orbit of the planets. They saw that the inner planets have a shorter distance to go, so they move more quickly because of the stronger pull of the Sun's gravity. Likewise, the outer planets have further to go, so they move more slowly because of their distance from the Sun and its gravitational pull.

Journal Prompt

Write a silly sentence designed to help you remember the order of the planets.
Example: My very excited mother just swam under nine porpoises.

Mercury	=	My
Venus	=	Very
Earth	=	Excited
Mars	=	Mother
Jupiter	=	Just
Saturn	=	Swam
Uranus	=	Under
Neptune	=	Nine
Pluto	=	Porpoises

Homework Idea

Have students go outside in the evening and stare up into the night sky. Encourage students to write down their thoughts.

Experiment 1: Food for Thought

- ***2 plums per student (Uranus and Neptune)***
- ***3 peppercorns per student (Mercury, Pluto, and Mars)***
- ***2 peas per student (Earth and Venus)***
- ***1 grapefruit per student (Jupiter)***
- ***1 large orange per student (Saturn)***

Try This!

Work with a partner to complete the experiment below. Record your findings on the **Food for Thought** Science Log.

Procedure:

1. Place your food items on a table.
2. Each item represents one of the nine planets. Discuss with a partner which planet is represented by each food item.
3. Discuss with your teacher which food items represent which planets. How close were your predictions? Place the planets in order on the table.
4. Record your observations on the Science Log.

What Happened?

How many Plutos would it take to fill the Neptune? How much larger is Jupiter than Earth? What else do you notice about how planets compare in size?

Experiment 2: The Planets Move Around the Sun

- ***piece of rope or sturdy string, 6 feet (1.8 m) long***
- ***sturdy post or pole***
- ***piece of rope or sturdy string, 10 feet (3 m) long***

Try This!

Work with a partner to complete the experiment below. Record your findings on **The Planets Move Around the Sun** Science Log.

Procedure:

1. Work with a partner. Each of you should tie one of the pieces of rope or string around the pole or post. Tie the string loosely, so that it turns around the pole as you move. Tie the longer rope higher up on the pole.

2. The post represents the Sun. The shorter string represents the orbit of an inner planet (such as Earth). The longer string represents the orbit of an outer planet.

3. Grab one string, while your partner grabs the other string. Step away from the pole until you are pulling the string taut. Begin walking around the pole, keeping the strings taut. The person representing an inner planet (pulling the shorter string) should move at a slightly quicker pace to represent the stronger gravitational pull of the Sun. The person representing an outer planet (pulling the longer string) should move at a normal pace.

What Happened?

What moved slower, the inner planet or the outer planet? What moved faster? Why do you think that happened?

Name ________________________

Science Log

Use this section to record your observations from the **Food for Thought** experiment.

Question: What happened and what did it show?

____

____

What we did:

____

____

How we did it:

____

____

Use this section to record your observations from **The Planets Move Around the Sun** experiment.

Question: Which planets move faster, inner or outer planets?

____

____

What we did:

____

____

How we did it:

____

____

WHAT IS THE SOLAR SYSTEM?

Activity 1: Show What You Know: The Solar System

Label this drawing using the words in the Word Bank.

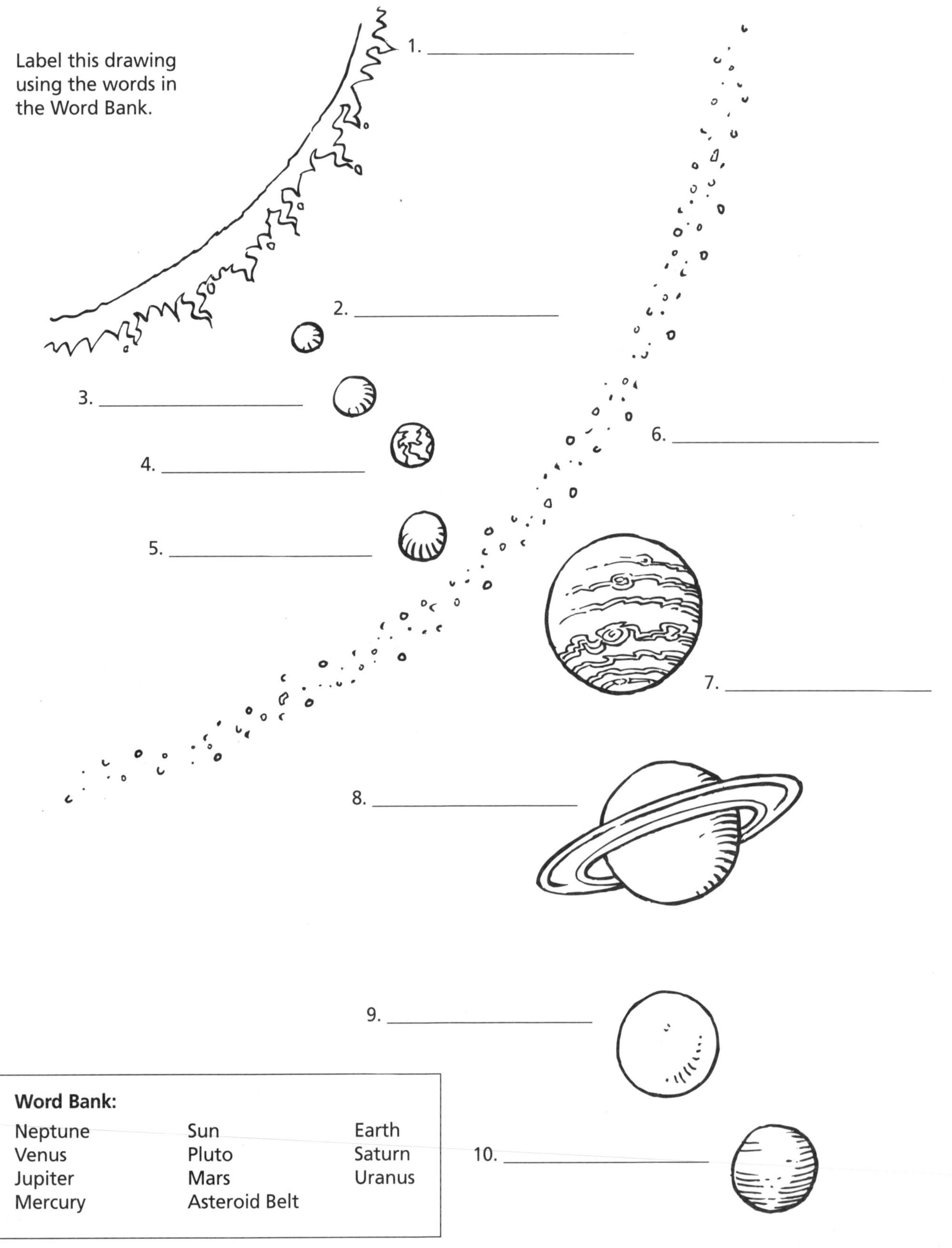

Word Bank:

Neptune	Sun	Earth
Venus	Pluto	Saturn
Jupiter	Mars	Uranus
Mercury	Asteroid Belt	

Lesson 2: The Sun

Use this page when you introduce the Sun to your students. The fun facts can be used to draw your students into the topic.

They'll Need to Know ...

The Sun is a star. It is the closest star to our planet and the largest, hottest, and brightest object in the Solar System. The Sun has more mass than all the other objects in our Solar System. Due to this great mass, the Sun exerts a powerful gravitational pull, and this holds all the planets (and some other celestial objects) in orbit around it.

The Sun mainly contains hydrogen gas. Atoms (or tiny particles) of hydrogen at the Sun's core may reach temperatures of up to 27 million degrees Fahrenheit. The higher the temperature, the more quickly atoms move. Fast-moving hydrogen atoms constantly crash into each other and form another gas known as helium. When this happens, energy is released. This energy warms the Sun and causes it to shine.

The Sun's light and energy are essential for our planet's survival. Plants use energy from sunlight to produce their own food. In turn, plants form the first link in the food chain. They feed many animals, which in turn feed other animals.

Prove It!

Remind students that even though they may find the Sun an interesting topic to study, they should never look directly at the Sun — the bright light could damage their eyes. You will find instructions for safely viewing the Sun in the **Safe "Sun" Glasses** experiment.

Experiment 1: Safe "Sun" Glasses
Teaching Notes:
Using a few simple materials, students were able to safely view the Sun without looking at it. Promote classroom discussion after this experiment by asking students what they saw and whether they saw any sunspots.

Experiment 2: A Plant Needs Sunshine
Teaching Notes:
In this experiment, students saw how important a role the Sun has in sustaining plants and other living things on our planet. You may want to use this experiment to lead into a discussion on photosynthesis.

Journal Prompt

Write about an activity you enjoy during summer, when days are warm and sunny.

Homework Idea

Have students answer the question: *Do you think artificial light is as beneficial as sunlight? Why or why not?*

- Sunspots are actually giant storms on the surface of the Sun. Solar flares are sudden bursts of hot, bright gases.
- The Sun uses up to 4 million tons of hydrogen per second, yet it still has enough hydrogen to exist for millions of years.
- One million Earths could fit inside the Sun.

Experiment 1: Safe "Sun" Glasses

- *cardboard*
- *white paper*
- *pen*
- *scissors*
- *tape*
- *binoculars*
- *stool or ladder*

Try This!

Work with a partner to complete the experiment below. Record your findings on the **Safe "Sun" Glasses** Science Log.

Procedure:

1. Place the binoculars' eyepieces on the cardboard and draw around them. Cut out the circles you have drawn to make two holes in the cardboard.
2. Push the binoculars' eyepieces though the holes in the cardboard. Tape the board in place, if needed.
3. Cover one of the large lenses (at the opposite end of the binoculars) with a piece of cardboard and tape it in place.
4. Go outside. Tape a sheet of white paper on a wall that is receiving plenty of sunlight. This paper is your viewing screen.
5. Stand about 3 feet (.9 m) away from the white paper screen. Hold the binoculars so sunlight shines through the one exposed large lens. Tilt and turn the binoculars until you see sunlight on your screen. You may need to reposition your screen if the sunlight shows up on another part of the wall.
6. Set the binoculars on a stool or a ladder once you get them in position. Closely observe the Sun's image on the white paper screen.
7. Place a piece of paper on your screen and draw what you see.

What Happened?

How did the binoculars allow you to view the Sun safely? Why do you think it is dangerous to look directly at the Sun? If you saw dark spots on the screen, what do you think they were?

Experiment 2: A Plant Needs Sunshine

- *2 identical potted plants*
- *scissors*
- *tape*
- *cardboard or construction paper*

Try This!

Work with a partner to complete the experiment below. Record your findings on the **A Plant Needs Sunshine** Science Log.

Procedure:

1. Cut and tape pieces of cardboard or construction paper to cover most of the leaves on one of the potted plants.
2. Place both plants in a sunny window, and care for them as you normally would for one full week.
3. At the end of the week, remove the cardboard from the plant's leaves.
4. Note your findings on your Science Log.

What Happened?

How did the leaves of the two plants compare to each other? Did different amounts of sunlight hit each plant's leaves? How is sunlight important to a plant?

Name ______________________________

Science Log

Use this section to record your observations after viewing the Sun in the **Safe "Sun" Glasses** experiment.

What we did: ______________________________

____

____

What I saw: ______________________________

____

My thoughts: ______________________________

____

____

Use this section to record your observations from the **A Plant Needs Sunshine** experiment.

What we did: ______________________________

____

____

What happened: ______________________________

____

____

Why this happened: ______________________________

____

____

Activity 1: Show What You Know: The Sun

Fill in the blank(s) for each sentence below. Use the Word Bank if you need help.

Word Bank:	
hydrogen	Gravity
star	eyes
Sunspots	mass
helium	life
Sun	
largest, hottest, and brightest	

1. The ________________ is our closest ________________.

2. The Sun is the ________________, ________________, ________________ object in the Solar System.

3. The Sun has more ________________ than any other object in our Solar System.

4. ________________ created by the Sun causes all the planets to orbit the Sun.

5. The Sun is made mostly of a gas called ________________.

6. Crashing atoms form another gas known as ________________.

7. The Sun's light and energy make ________________ possible on Earth.

8. ________________ are actually giant storms on the Sun's surface.

9. Looking directly at the Sun can cause severe damage to your ________________.

Lesson 3: Tilted, Rotating & Revolving

Use this page when you introduce the concepts of Rotating and Revolving to your students. The fun facts can be used to draw your students into the topic.

They'll Need to Know ...

Each of the nine planets moves in two distinct ways.

Every planet spins, or rotates, on its own axis. An axis is an imaginary line that runs through the center of a planet, from one pole to the other. The time it takes for a planet to complete one full rotation around its axis is known as a day. Wherever a planet receives light from the Sun, it is day. Wherever a planet is not receiving light, it is night. Earth spins counterclockwise once around its axis every 24 hours, therefore this is the length of an Earth day. Jupiter experiences the shortest day of all the planets — a little less than 10 hours. Venus, on the other hand, rotates very slowly. One day on Venus equals 243 Earth days!

Earth does not spin straight up and down on its axis. Instead, it spins at a slight tilt. Other planets spin on a tilt, too. In fact, Uranus spins sideways! Scientists believe it may have been struck by another large object in the past.

The planets also revolve, or circle, around the Sun. The path each planet follows around the Sun is called its orbit. The time required for a planet to complete its orbit is called a year. The farther a planet is located from the Sun, the longer it takes to complete its orbit. One year on Earth measures about 365 and one-fourth days. Pluto, the farthest planet from the Sun, takes 247 Earth years to complete its orbit.

As Earth revolves around the Sun, it sometimes tilts toward the Sun and sometimes tilts away from the Sun. This is what causes seasons. For about six months, the North Pole is tilted toward the Sun. This provides the northern region, called the Northern Hemisphere, with warmer temperatures during spring and summer. At the same time, the southern region, called the Southern Hemisphere, experiences cooler temperatures, and fall and winter seasons take place there.

Prove It!

The relationship the Sun has to Earth's seasons may be a difficult concept to understand. Therefore, allow for plenty of time in completing the experiment and activity on the following pages. Promote classroom discussion of the experiment by asking students for their questions upon completion. You may also look to their Science Logs for any additional questions that need covering. The activity **Make a Sundial** contains instructions for creating one of the first instruments used for telling time. You will need to help students position the sundial so it points North and South. Be sure students do not move their sundials as they trace shadows from them. If the day after this activity is not expected to be sunny, mark the spot where you placed the sundial and set it there on the next sunny day. Note: this sundial is not accurate after one month, due to the fact that Earth is revolving.

Experiment 1: Understanding Day, Night, and the Seasons Teaching Notes:
In this experiment, students saw the connection between Earth's movement in space and natural cycles such as day and night and the seasons.

Journal Prompt

How would life be different for you if you lived in Australia? Write about your favorite season.

Homework Idea

Have students discuss time zones with an adult and then write an answer to the question, *How do time zones relate to the study we completed today?*

- The Moon also has day and night periods. However, the Moon spins much more slowly than Earth, so a complete day and night is actually two weeks long!
- The equator receives similar amounts of sunlight all year long, so it does not experience four seasons.
- If Earth did not spin on a tilt, there would not be any seasons.

Experiment 1: Understanding Day, Night, and Seasons

- *lamp with shade removed or flashlight*
- *globe*
- *small paper doll or animal*
- *tape*

Try This!

Work with a partner to complete the experiment below. Record your findings on the **Understanding Day, Night, and Seasons** Science Log.

Procedure:

1. Stand the lamp or flashlight on a large table.
2. Tape the paper doll to the globe on any country or your own.
3. Place the globe at least 4 feet (1.2 m) away from the light source, which represents the Sun.
4. Turn on the light source, and turn off the lights.
5. Slowly turn the globe to the left (counterclockwise) until the Earth has made one full turn. This full rotation represents one full day. As you do this, watch the doll and note when the doll would be asleep or awake (when night or day would occur). Also record which countries are experiencing night when the doll's country is experiencing day.
6. Next, hold the globe by its base and walk it in a circle around the light source. Be sure not to change the globe's position. It should always tilt the same way. Observe how light strikes the doll's country differently as Earth circles the Sun.

What Happened?

How does Earth's rotation cause day and night? How does Earth's revolution cause seasons? When did the doll's country experience summer?

Name ______________________

Science Log

Use this sheet to record your observations from the **Understanding Day, Night, and the Seasons** experiment.

Question: What did I learn about Earth?

____

____

____

What we used and what the materials represented:

____

____

____

What we did:

____

____

____

What I learned about Earth's days and seasons:

____

____

____

Activity 1: Make a Sundial

Complete the activity below to make an ancient tool for tracking time.

- ***wood board, about 6 inches (15.2 cm) wide x 9 inches (22.9 cm) long***
- ***small piece of modeling clay***
- ***watch or clock***
- ***pencil, about 7 inches (17.8 cm) to 8 inches (20.3 cm) long***
- ***triangle or carpenter's square***
- ***magnetic compass***

Procedure:

1. Draw a straight line through the middle of the board and parallel to one side.
2. Place a clump of modeling clay in the middle of the line you drew.
3. Push the eraser end of the pencil into the clay, so that the pencil stands upright.
4. Using a carpenter's square, make sure that the pencil is perpendicular to the board.
5. Put your sundial to work! Here's how:
   - Place the sundial outside in direct sunlight on a sunny morning.
   - Have your teacher help you use a compass to place your sundial so it points North and South.
   - Observe where the pencil makes a shadow on the sundial every hour on the hour.
   - Trace the shadow on the board.
   - Write the hour above the mark, and leave the sundial in position.
   - Use it the next day to tell time.

Name ______________________________

Activity 2: Show What You Know: Tilted, Rotating & Revolving

Label each sentence T for true or F for false using what you know about Earth's movement in space.

_______1. Day and night are caused by the Earth revolving around the Sun.

_______2. A sundial is a tool used to tell time.

_______3. Earth's tilt causes the seasons.

_______4. When the North Pole points towards the Sun, the Northern Hemisphere experiences warmer temperatures of spring or summer.

_______5. All planets have the same length of day.

_______6. Each planet's year is a different length.

_______7. A planet's path around the Sun is called an orbit.

_______8. An axis is an imaginary line that runs through the center of a planet.

_______9. A day is one full rotation around a planet's axis.

_______10. A year on Earth is 24 days long.

Lesson 4: Inner Planets: Mercury, Venus, Earth & Mars

Use this page when you introduce the planets Mercury, Venus, Earth, and Mars to your students. The fun facts can be used to draw your students into the topic.

They'll Need to Know ...

During lessons 4, 5, and 6, you will introduce students to each of the nine planets. The information will be presented in a data file format. Share the information with your students, and enhance the lessons with actual photos and drawings of each planet. In addition, always take time to refer to the planet on a classroom map of the Solar System. Finally, use the **Planet Fact File** (page 44) to allow students to record important data on each planet. You will need to make nine copies for each student, one for each planet.

Lesson 4 focuses on the inner planets. These planets are closest to the Sun. They are made of rock and metals, and they are heavy and dense. The inner planets have only a thin layer of atmosphere, and they are the warmest planets.

Statistics	**Mercury**	**Venus**	**Earth**	**Mars**
Position from Sun	first	second	third	fourth
Distance from Sun (in miles)	36 million	67 million	93 million	142 million
Diameter (in miles)	3,050	7,520	7,930	4,220
Day Length (in Earth days)	59	243	1	1.03
Year Length (in Earth days)	88	225	365	687
Number of Moons	none	none	1	2
Rings	no	no	no	no

- Mercury is pitted with craters. The craters were more than likely created by meteors.
- Mercury has virtually no atmosphere because it is so hot — temperatures can reach more than 800 degrees Fahrenheit.
- Venus and Earth are similar in size.
- Venus' atmosphere is filled with deadly carbon dioxide, and its clouds are made of sulfuric acid.
- Venus rotates on its axis in a clockwise direction (opposite from all the other planets), so if you lived on Venus, the Sun would rise in the west and set in the east!
- Venus is easy to spot in the night sky because it shines very brightly.

Lesson 4: Inner Planets: Mercury, Venus, Earth & Mars (continued)

They'll Need to Know ...

The inner planets are similar in make up. The first experiment simulates what happens when meteoroids hit an inner planet. NOTE: A meteoroid that hits Earth is called a meteorite. See page 59 for an explanation of the difference. Be sure to explain how other inner planets may react the same way given the similar makeup of the crust. You may also want to introduce atmosphere at this time and how the inner planets have different amounts of atmosphere that help protect them from meteoroids. Then, make copies of the **Planet Fact File** on page 44 for each planet studied.

Experiment 1: Create Mercury's Craters
Teaching Notes:
Mercury's surface has many craters. The fact that this planet doesn't have an atmosphere, and therefore no weather, means there's nothing to help wash away the craters or burn up the meteoroids that scientists believe slammed into the planet over and over again to create the craters. In this experiment, students were able to see how meteoroids can change the surface of a planet.

Journal Prompt

Imagine life exists on Mars, and describe what that life might be like. Would you ever want to be part of a NASA mission to Mars? Why or why not?

Homework Suggestion

Have students search online for great information on Mars and answer the following question: *Why do scientists believe Mars is the only other planet in the Solar System able to sustain life?*

- Water covers more than 70% of Earth's surface.
- Ice covers Earth's North Pole and South Pole.
- Earth is the only planet in the Solar System known to sustain life.
- Mars is called the "Red Planet" because a layer of soft, red iron oxide covers it.
- Mars is the planet that's most like Earth.
- Mars has the largest volcano in the Solar System.
- Mars has a series of channels on its surface. These might have been created when it was warmer and had rivers of water. The channels are now dry.
- There is no evidence of life on Mars.

Experiment 1: Create Mercury's Craters

- ***drop cloth or newspapers***
- ***glass bowl, as large as possible***
- ***flour, loose dirt, or sand***
- ***small stool or chair***
- ***marbles***
- ***rocks***
- ***small ball***
- ***small building blocks***

Try This!

Work with a partner to complete the experiment below. Record your findings on the **Create Mercury's Craters** Science Log.

Procedure:

1. Spread the drop cloth or newspapers on the floor. Be sure to cover a large area.
2. Place the flour, loose dirt, or sand in the bowl. Place the bowl in the center of the drop cloth.
3. Take turns standing on a small stool or chair and dropping the marbles, rocks, and other objects into the bowl.
4. Take a break every once in a while, to observe the "craters" created by the objects.

What Happened?

What does the material in the bowl represent? What do the falling objects represent? How do the shapes made by the objects compare to each other?

Name ______________________

Science Log

Use this sheet to record your observations from the **Create Mercury's Crater's** experiment.

Question: What do scientists believe caused Mercury's craters? ______________________

____

____

Materials I used: ______________________

____

____

What I did: ______________________

____

____

What happened: ______________________

____

____

Illustrate your experience:

Activity 1: Show What You Know: Planet Fact File

Planet's Name: ______________________

Position from the Sun: ______________________

Distance from the Sun: ______________________

Diameter: ______________________

Rings?: ______________________

Day Length: ______________________

Year Length: ______________________

Number of Moons: ______________________

Did You Know? ______________________

____

____

____

____

____

Draw a picture of the planet.
Label any interesting features.

Lesson 5: Outer Planets Part 1: Jupiter & Saturn

Use this page when you introduce Jupiter and Saturn to your students. The fun facts can be used to draw your students into the topic.

They'll Need to Know ...

The outer planets are colder and darker than the inner planets. They are gaseous and usually contain liquid or ice. Share the information below on Jupiter and Saturn with your students, and enhance the lessons with actual photos and drawings of each planet. In addition, always take time to refer to the planet on a classroom map of the Solar System.

Statistics	Jupiter
Position from Sun	fifth
Distance from Sun (in miles)	484 million
Diameter (in miles)	89,000
Day Length (in Earth time)	9 hrs., 53 min.
Year Length (in Earth years)	11.9
Number of Moons	16
Rings	yes

- Jupiter is the largest planet in the Solar System and is sometimes called "The Giant."
- Jupiter is larger than all the other planets.
- The dark bands around Jupiter are called belts and the light bands are called zones.
- The great red spot on Jupiter is actually a storm on its surface. This storm stretches more than 25,000 miles (40,225 km)!
- In the night sky, Jupiter appears to be a very bright star that doesn't twinkle.

Lesson 5: Outer Planets Part 1: Jupiter & Saturn (continued)

Prove It!

Jupiter is notorious for its storms. The great red spot that's made Jupiter famous is actually a huge storm. Scientists aren't sure what causes the storms, but they can churn streams of gas that move faster than the planet is actually rotating! Before you have students complete the activity on the page that follows, make sure that students know how the inner planets differ from the outer planets. Then use the experiment on page 47 to have students create a Jupiter storm. Upon completion, photocopy the **Planet Fact File** on page 44 to allow students to record important data on Jupiter and Saturn.

Experiment 1: Create a Jupiter Storm
Teaching Notes:
With this experiment, students modeled the way that storms form on the surface of Jupiter. Be sure to explain to students that Jupiter's gaseous makeup causes its storms to be different from storms on Earth.

Statistics	Saturn
Position from Sun	sixth
Distance from Sun (in miles)	885 million
Diameter (in miles)	75,000
Day Length (in Earth time)	10 hrs., 40 min.
Year Length (in Earth years)	29.5
Number of Moons	21
Rings	yes

Journal Prompt

Imagine telling a creature from Jupiter or Saturn how life on Earth is different from its planet. Write about your thoughts.

Homework Idea

Have students use art materials to create a picture of Jupiter or Saturn. Encourage students to share their pictures with the class.

- Saturn's rings extend over 171,000 miles (275,139 km).
- Saturn's rings are made of frozen stones. These stones can be very large, sometimes as big as a car!
- Saturn was named after the Roman god of farming.
- It is difficult to see Saturn in the night sky. However, it can be viewed with a telescope.
- Although Saturn's diameter is nearly 10 times larger than Earth's diameter, it would float if placed in water and Earth would sink. This is because Saturn is mainly made of gases while Earth is made up of rocks.

Experiment 1: Create a Jupiter Storm

- *large glass bowl*
- *milk*
- *measuring cup*
- *eye dropper*
- *yellow food coloring*
- *red food coloring*
- *dishwashing liquid*

Try This!

Work with a partner to complete the experiment below. Record your findings on the **Create a Jupiter Storm** Science Log.

Procedure:

1. Pour two cups of milk into the bowl.
2. Add two drops of red food coloring to the milk. Add two drops of yellow food coloring.
3. Spin the bowl gently. The milk should swirl around, but the colors should not mix.
4. Place one to two drops of dishwashing liquid on top of the drops of food coloring.
5. Spin the bowl gently and observe what happens.

What Happened?

How is your milk/food coloring/dishwashing liquid model similar to conditions on Jupiter? What happened when you added dishwashing liquid to your model?

Name ______________________

Science Log

Use this sheet to record your observations and work from the **Create a Jupiter Storm** experiment.

Question: What are storms like on Jupiter? ______________________

____

____

What I did: ______________________

____

____

What I used: ______________________

____

____

What happened: ______________________

____

____

Illustrate what happened:

Lesson 6: Outer Planets Part 2: Uranus, Neptune & Pluto

Use this page when you introduce the planets Uranus, Neptune, and Pluto to your students. The fun facts can be used to draw your students into the topic.

They'll Need to Know ...

Share the information on these two pages with your students, and enhance the lessons with actual photos and drawings of each planet. In addition, always take time to refer to the planet on a classroom map of the Solar System.

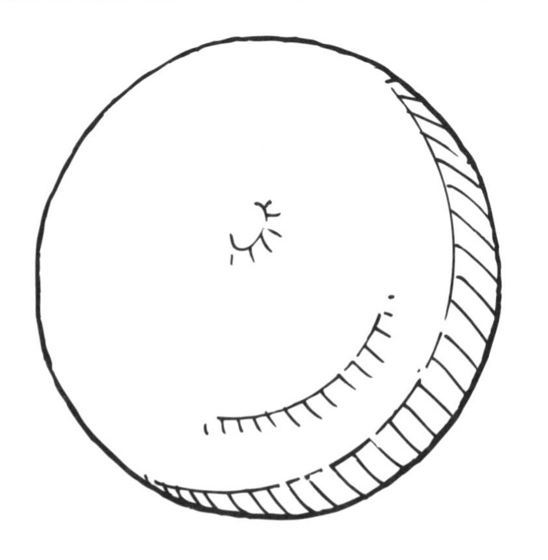

Statistics	Uranus
Position from Sun	seventh
Distance from Sun (in miles)	783 million
Diameter (in miles)	31,764
Day Length (in Earth time)	17 hours
Year Length (in Earth years)	84
Number of Moons	15
Rings	yes

- Uranus rotates differently than other planets; it seems to be on its side. Some scientists think a meteoroid may have hit the planet and knocked it that way.
- Methane gas in Uranus' atmosphere gives it a bluish tint.
- There are very faint gray rings surrounding Uranus. Some scientists think the rings are made of graphite.
- Uranus is difficult to see without the aid of a telescope.
- Uranus was named after the father of Saturn, a Greek god.
- Neptune has no solid ground.
- Neptune has storms that come and go.
- Neptune can only be seen with a telescope.
- The winds on Neptune's surface can reach amazingly high speeds.
- For a short period every 232 years, Neptune is the farthest planet in our Solar System. During this time, Pluto's unusually shaped orbit brings it in front of Neptune, placing it closer to the Sun. This will occur again in 2231.
- Pluto is a combination of rock and ice. It is the only outer planet that contains rock.
- Scientists can only view Pluto with a high-powered telescope.
- Pluto was discovered in the 1930s. There may be another planet further beyond it called Planet X!

Lesson 6: Outer Planets Part 2: Uranus, Neptune & Pluto (continued)

Prove It!

The experiment on page 51 helps explain how Uranus's rings were discovered and why the rings were so hard to spot. After students complete this experiment and have a better understanding of all of the outer planets, use the **Planet Fact File** on page 44 to allow students to record important data on the remaining three planets.

Experiment 1: Discover Uranus's Rings
Teaching Notes:
The fact that Uranus has rings is relatively new scientific information. In 1977, Uranus passed in front of a very bright star. As Uranus passed the star, the light flickered. Scientists were able to determine that Uranus's rings caused the flickering! In this experiment, students reconstructed the situation that allowed scientists to identify Uranus's rings. Be sure to point this out before starting the experiment.

Journal Prompt

If you could visit one of the planets studied today, which planet would you choose and why?

Homework Idea

Have students write a letter to an astronaut who is going on a voyage to Pluto and include any questions they want answered.

Statistics	Neptune	Pluto
Position from Sun	eighth	ninth
Distance from Sun (in miles)	2,790 million	3,660 million
Diameter (in miles)	30,800	1,620
Day Length (in Earth time)	16 hours	6 days, 9 hrs.
Year Length (in Earth days)	165	248
Number of Moons	8	1
Rings	yes	no

Experiment 1: Discover Uranus's Rings

- *rectangular block of Styrofoam™, at least 5 inches (12.7 cm) wide x 15 inches (38 cm) long*
- *black paint*
- *flashlight*
- *ruler*
- *10 to 15 pencils*

Try This!

Work with a partner to complete the experiment below. Record your findings on the **Discover Uranus's Rings** Science Log.

Procedure:

1. Paint the Styrofoam™ with black paint.
2. Paint the pencils black.
3. Stick the pencils into the Styrofoam™ pointed side down. Create a straight row of pencils that runs the length of the Styrofoam™ block. Make sure the pencils stand straight up. The pencils represent Uranus' rings.
4. Have a partner hold the flashlight approximately 3 feet (.9 m) in front of the Styrofoam™ block. Turn on the flashlight. This represents a bright star. Turn off the lights.
5. The other partner should sit near the row of pencils, on the side opposite from the flashlight, and watch as the flashlight moves from side to side.

What Happened?

What causes the flashlight beam to flicker? How do you think this relates to how Uranus's rings interact with light shining from a star? How do shadows created by the rings help scientists know they are present around Uranus?

Name ______________________

Science Log

Use this sheet to record your experience while constructing your model in the **Discover Uranus's Rings** experiment.

Question: Why was it so difficult for scientists to know if Uranus had rings?

____

____

____

Materials we used: ______________________

____

____

What we did: ______________________

____

____

Draw and label a sketch of your model:

Lesson 7: The Moon

Use this page when you introduce the Moon to your students. The fun facts can be used to draw your students into the topic.

They'll Need to Know ...

The Moon is a satellite of the Earth — it revolves around the Earth due to gravity from Earth. The Moon does not produce its own light. Instead, it reflects light from the Sun.

Another interesting fact about the Moon is that crashing meteoroids have created multiple craters on its surface, which gives it a dimpled appearance. These patterns of craters produced a face-like image on the Moon, giving rise to the phrase "The Man on the Moon." The moon also has no atmosphere, and its gravity is one-sixth that of Earth's gravity. As a result, a person would weigh one-sixth his or her Earth weight if he or she was standing on the Moon.

The Moon is the only other celestial body that humans have explored in person. *Apollo 11* landed on the Moon in 1969. Neil Armstrong, an astronaut, was the first man to walk on the Moon.

The Moon seems to change appearance in our night sky. It doesn't actually change. Rather, the amount of sunlight reflected off the side of the Moon that we are able to view changes as it revolves around the Earth. The different shapes we see are called phases.

Earth always has a shadow that's created by the Sun and on very rare occasions, the Sun and the Earth line up when the Moon passes through the shadow. A solar eclipse occurs when the Moon passes between the Earth and the Sun. A lunar eclipse occurs when the Moon passes through Earth's shadow.

Prove It!

Using the information on the experiment pages that follow, students can simulate the Moon's phases and a solar eclipse. Before you have students conduct the second experiment, make sure that students understand how a solar eclipse occurs (i.e.,when the Moon passes between the Sun and Earth).

Experiment 1: Understanding the Phases of the Moon Teaching Notes:
In this experiment, students revolved their Moon around Earth near a light source so they could see the different light patterns appear on the Moon. This experiment will help students understand the phases of the Moon.

Experiment 2: Simulating a Solar Eclipse Teaching Notes:
In this experiment, students were able to model a solar eclipse (i.e., how the Moon can block the Sun's light from hitting Earth). Be sure to remind students that though the Moon is many times smaller than the Sun, it is closer to the Earth and therefore appears larger.

Journal Prompt

Imagine being able to walk on the Moon. Describe what it would be like as you moved around and looked through space to see Earth.

Homework Idea

Have students go outside at night and draw the Moon they see. Give students the Moon Phases chart from page 57 to determine what phase it is in. Encourage students to draw the Moon on the same night each week and chart the phases.

- The oldest Moon rocks may date back more than 4 billion years.
- A person who weighs 100 pounds on earth would weigh only 17 pounds on the Moon.
- Although the Moon looks about the same size as the Sun, the Moon is actually much smaller. It only appears larger because it is much closer to Earth.

Experiment 1: Understanding the Phases of the Moon

- *flashlight*
- *golf ball or Ping-Pong ball*
- *tennis ball or softball*
- *aluminum foil*

Try This!

Work with a partner to complete the experiment below. Record your findings on the **Understanding the Phases of the Moon** Science Log.

Procedure:

1. Cover the small ball with aluminum foil. This will be your Moon. The large ball will represent Earth.
2. Place your Earth and Moon on a table, about 8 inches (20.3 cm) apart.
3. Turn on the flashlight. Place it about 3 feet (.9 m) away from your Earth and Moon. In this activity, the flashlight represents the Sun.
4. Turn off the classroom lights.
5. Slowly move the Moon around Earth. Be sure to keep the Moon at the same distance from Earth as you take it through its orbit. Note the way the light appears on the Moon. This is similar to how the Moon's phases appear to you each night over a complete lunar cycle.

What Happened?

Imagine being a person on Earth (the large ball) looking at the Moon (the small ball). When would you see the most light? When would you see the least light? How does the Moon's movement cause you to see different amounts of light?

Experiment 2: Simulating a Solar Eclipse

- ***lamp with shade removed***
- ***tennis ball***

Try This!

Use simple materials to model a solar eclipse. Then, record your findings on the **Simulating a Solar Eclipse** Science Log.

Procedure:

1. Turn on the lamp.
2. Stand about 10 feet (3 m) away from the lamp, and face it. Your body is Earth.
3. Hold the tennis ball (the Moon) at arm's length, toward the lamp. Try to block the view of the lamp (the Sun) with the Moon.
4. Take this experiment a step further. Take a quarter and go outside. Locate a large building in the distance. Try to block out the building with the quarter.

What Happened?

Where was the Moon in your model compared to the Sun and Earth? Did the same thing happen with the quarter?

Name ______________________

Science Log

Use this section to record your observations from the **Understanding the Phases of the Moon** experiment.

What we used: ______________________

____

What we did: ______________________

____

What happened: ______________________

____

Why it happened: ______________________

____

Use this section to record your observations from the **Simulating a Solar Eclipse** experiment.

What we used: ______________________

____

What we did: ______________________

____

What happened: ______________________

____

Why it happened: ______________________

____

Activity 1: Show What You Know: The Moon

Complete each sentence below with the word that fills in the blank. Then, use that word to fill in the crossword puzzle.

Across

1. ________________ between Earth and the Moon keeps the Moon in its orbit around our planet.
2. Changes in the Moon's appearance are also called ________________.
3. A ________________ eclipse occurs when the Moon passes through Earth's shadow.
4. Neil ________________ was the first man to walk on the Moon.
5. ______________________ crashing into the Moon are thought to have caused its craters.
6. The Moon reflects light from the ________________.

Down

7. A full lunar cycle lasts approximately 29 ________________.
8. A ________________ eclipse occurs when the Moon passes between the Sun and Earth.
9. The Moon's surface has many ________________.

First Quarter
Waxing Gibbous
Waxing Crescent
Full Moon
New Moon
Sun
Waning Gibbous
Waning Crescent
Third Quarter

Lesson 8: Our Night Sky

Use this page when you introduce Stars, Constellations, and other elements of our Night Sky to your students. The fun facts can be used to draw your students into the topic.

They'll Need to Know ...

An astronomer is a scientist who studies the sky and its objects. On a clear night as many as 5,000 stars are visible to the naked eye. Stars are hot, glowing balls of gas. They differ in mass, size, temperature, and brightness (also called luminosity). Stars form out of clouds of gas and dust in space. After billions of years, a star may explode or collapse in on itself.

Stars are very luminous. This means they create and give off their own light. The light is a byproduct of the constantly burning gases that make up a star — helium and hydrogen. Heat is another byproduct of this process.

The Sun is our nearest star. We can feel its heat and use the light it provides. Most of the other stars we see are very far away. A star's distance from the Earth is measured in light years. A light year is equal to the distance that light travels in one year.

A constellation is a group of stars that, when mapped out and connected with lines, appear to have a recognizable shape. Most of the constellations were identified and named hundreds of years ago. Their names typically refer to the zodiac or famous Greek myths. Star charts are a fun way to view the constellations. The easiest star grouping to identify is the Big Dipper, which is part of the bear constellation Ursa Major.

A large grouping of stars is called a galaxy. There are many galaxies in our universe. Our Solar System is part of the Milky Way galaxy. On a very clear night, you may be able to see a part of the Milky Way. It appears as a shimmering band across the sky.

Other Members of the Night Sky

Comets

A comet's main ingredients are ice, gas, and dust. According to astronomers, comets are made from material left over from when the Solar System first formed. A comet orbits the Sun, just as planets do. The solid part of a comet is called the nucleus. It is mainly solid gas, and it measures just a few miles across. When a comet's orbit brings it near the Sun, the Sun's heat begins to melt the frozen gas in the nucleus. This creates exploding jets of gas, which form a giant ball of gas and dust around the nucleus. The giant ball is called the coma.

As the comet travels, it grows tails of gas and dust. These tails may be blue, yellow, or white, and may be as long as 60 million miles (97 million km)! At times, comets can be seen from Earth. They look like bright streaks of light in the sky.

- The coolest, dimmest stars are reddish. Warmer, brighter stars are yellow or white. The hottest and brightest stars are blue-white. The Sun is a medium-size white star, and it is the closest star to Earth. Still, the Sun has an average distance from Earth of about 93 million miles (150 million km).
- The constellation Cassiopeia is named after a queen in a Greek myth. In the myth, she is often seen sitting on her throne with a mirror in her hand. During most of the year, the constellation named in her honor appears to be turned upside down. Perhaps sitting in this position is punishment for all her bragging and self-centered behavior.

Lesson 8: Our Night Sky (continued)

Meteoroids

Meteoroids are solid pieces of rock and dust that break off from comets or colliding asteroids. The craters on Mercury and our Moon are thought to have been caused by crashing meteoroids. Depending on where they are in the Solar System, meteoroids are scientifically referred to in three different ways:

- meteoroid — small pieces from larger bodies, such as planets and asteroids
- meteor — the streak of light that occurs when a meteoroid enters Earth's atmosphere
- meteorite — a meteoroid that reaches Earth without burning up in the atmosphere

Asteroids

Asteroids are bodies smaller than planets that orbit the Sun. Asteroids are often called minor planets. Most of the 6,000 asteroids astronomers have located and named are part of the Asteroid Belt. This concentration of asteroids lies between Mars and Jupiter and contains thousands of asteroids of different sizes and makeup. The largest known asteroid is called Ceres. It measures 580 miles (933 km) across.

Black Holes

Black holes occur when older stars use up all their gases and die. They collapse inward, shrinking until there is nothing left but a point in space. A black hole has a very strong gravitational pull. Anything that comes near it can be sucked into the hole and never be seen again!

Prove It!

The **Make a Constellation** experiment on page 60 will help you create a classroom observatory. Before doing this experiment, you will need to gather clean, empty soup cans and cut off the ends with a can opener. Make sure there are no rough edges around the cans.

Experiment 1: Make a Constellation
Teaching Notes:
In this experiment, students created miniature versions of the machines planetariums use to show others the night sky. Encourage students to further their investigation of constellations by having them research their favorite and create another constellation can, this time by mapping out their constellation. Students may want to select their zodiac sign or a constellation that is easy to view in your area.

Journal Prompt

If you discovered a new constellation, what would you name it and how would it appear?

Homework Idea

Encourage students to use a field guide to identify stars, constellations, planets, and other objects in the night sky above their home.

Experiment 1: Make a Constellation

- ***Mini-Constellations sheet***
- ***scissors***
- ***soup can with ends removed***
- ***tape***
- ***black construction paper***
- ***pin or sharp pencil point***
- ***flashlights***

Try This!

Work with a partner to complete the experiment below.

Procedure:

1. Cut one constellation of your choice from the **Mini-Constellations** sheet.
2. Cut out a black circle that's a little larger than the base of the soup can.
3. Tape the black circle over one end of the can.
4. Hold the constellation pattern securely against the black paper. Use a pin or a pencil point to poke holes through each dot on the constellation pattern. In this way, you will transfer the pattern onto the black paper.
5. Your teacher will give you a flashlight. Place the flashlight inside the open end of the can, and aim it at a wall or the ceiling. When your teacher darkens the room, turn on your flashlight. What do you see?

What Happened?

What is the name of the constellation you chose? When you looked at the pinpoints of light, did the constellation live up to its name? How did your constellation receive its name?

Mini-Constellations

Use this page for the **Make a Constellation** experiment.

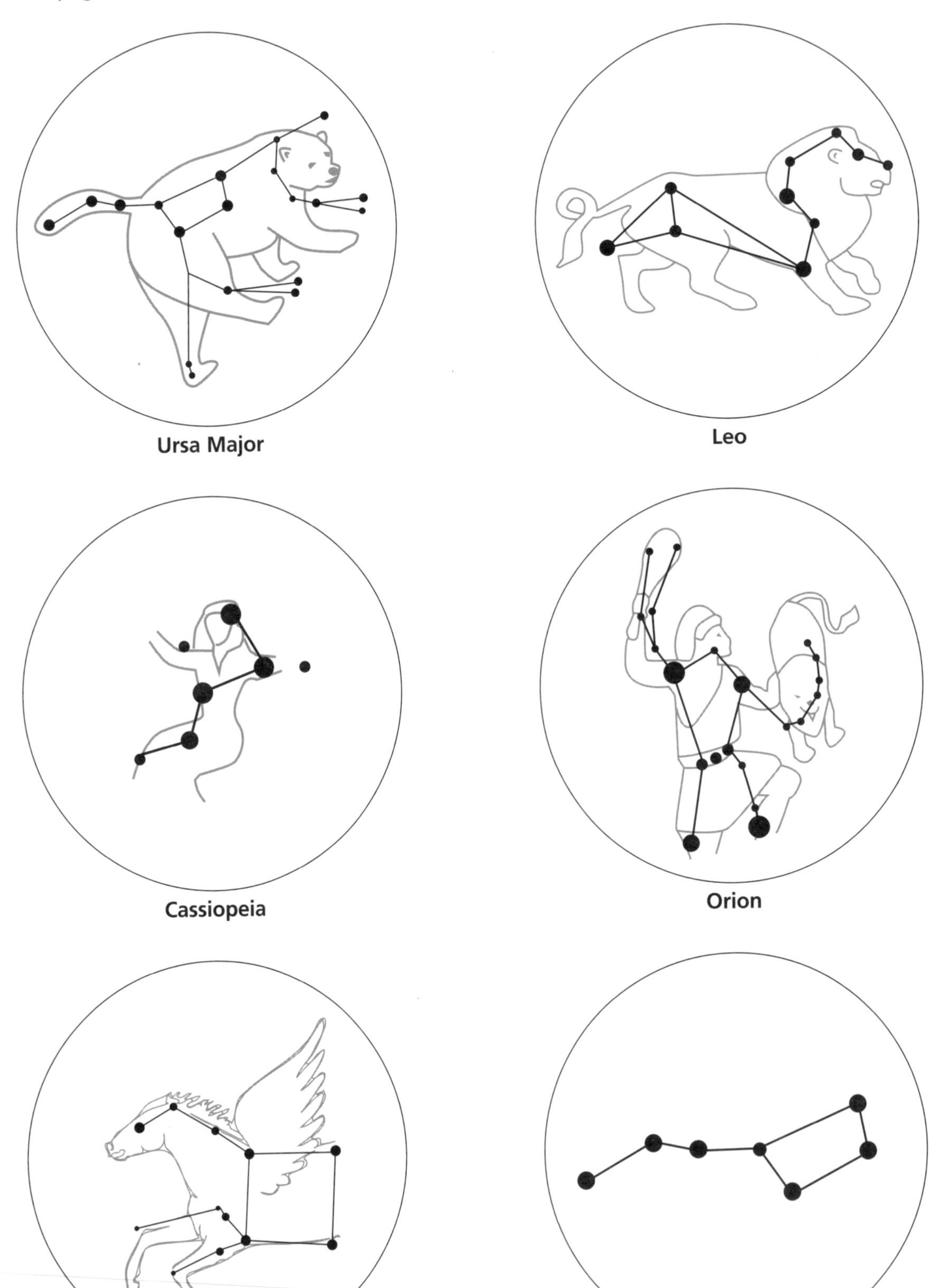

Big Dipper
(this is a star grouping that's part of the constellation Ursa Major. Can you find the Big Dipper in Ursa Major above?)

Activity 1: Show What You Know: Our Night Sky

Write a definition for each of the terms listed below.

1. asteroids: ______________________________

____

2. astronomer: ______________________________

____

3. Big Dipper: ______________________________

____

4. black hole: ______________________________

____

5. comet: ______________________________

____

6. constellation: ______________________________

____

7. galaxy: ______________________________

____

8. luminous: ______________________________

____

9. star: ______________________________

____

Lesson 9: History of Astronomy

Use this page when you introduce the History of Astronomy to your students. The fun facts can be used to draw your students into the topic.

They'll Need to Know ...

Spend this lesson reviewing a brief history of astronomy. Discuss how long curious scientists have been looking skyward. You may wish to photocopy this page and page 64 for students to use as a reference.

Aristotle (384-322 BC)
Aristotle believed that Earth was the center of the universe. He also stated that Earth was made up of only four elements: earth, water, air, and fire. He thought that the other celestial bodies such as the Sun, Moon, and stars, were made of a fifth element called ether.

Ptolemy (85-165 AD)
Ptolemy was a Greek astronomer. He combined his ideas with those of Aristotle and created his own model of the universe. Like Aristotle, Ptolemy also believed that Earth was the center of the universe, and all other heavenly bodies circled it. This view remained intact for over 1,400 years until the time of Copernicus.

Hypatia (370-415 AD)
Hypatia was an Egyptian math and astronomy teacher and is remembered as one of the first female astronomers. In addition to teaching math and astronomy, Hypatia invented several tools relating to astronomy and the Earth sciences.

Copernicus (1473-1543 AD)
Nicholas Copernicus was a Polish astronomer. He was the first scientist to introduce a model of the Solar System that centered around the Sun. He believed that all the planets, including Earth, moved in orbits around the Sun.

Tycho Brahe (1546-1601 AD)
Tycho Brahe was a Danish astronomer. He is best known for his observations of stars, comets, and planets in the night sky. Tycho also built the first observatory. Tycho's records were later used by Kepler to describe the orbits of planets.

- Galileo's work suggested the Solar System had flaws. This idea of the heavens' imperfection offended the Roman Catholic Church. He was sentenced to house arrest for the final years of his life.
- Pluto is considered a recent discovery in astronomy. Clyde Tombaugh discovered the distant planet in 1930.

Lesson 9: History of Astronomy (continued)

Johan Kepler (1571-1630 AD)
Johan Kepler was a German astronomer. Kepler used Brahe's records to show that planetary orbits are not circular, but elliptical (or shaped like an egg). He was able to prove this theory about planetary motion. As a result, he helped prove that the Copernican model of the universe was correct.

Galileo Galilei (1564-1642 AD)
Galileo Galilei was an Italian astronomer and physicist. He was the first scientist to use a telescope to study the night sky and Jupiter's moons. Galileo was also one of the first scientists to report that there were craters on the Moon and spots on the Sun.

Sir Isaac Newton (1642-1727 AD)
Sir Isaac Newton was an English scientist and mathematician. He was one of the most notable scholars of his time. Newton developed several laws that help us understand the movement of objects. As the story goes, Newton was inspired to create his theory of gravity after he watched an apple fall to the ground.

Caroline Herschel (1750-1848 AD)
Caroline Herschel was a German astronomer. She and her brother, William Herschel, discovered three new nebulae (clouds of space dust) and eight comets. She also wrote books on astronomy.

Prove It!

Explain to students that Newton's Third Law states, "For every action, there is an equal and opposite reaction." Also explain to students that it is at work when two people playing football bump into each other and they both fall to the ground. Then, have students conduct the experiment on page 65.

Experiment 1: Newton's Third Law and Rockets
Teaching Notes:
Newton's Third Law states, "For every action, there is an equal and opposite reaction." Rocket engines work in a similar way. Students saw this in action with this experiment. The air shot out of the balloon in one direction, pushing the balloon in the opposite direction.

Journal Prompt

If you could meet one astronomer, who would you like to meet and why?

Homework Idea

Have students write an essay describing how they think the scientists they studied today must have been viewed by fellow citizens of their time.

Experiment 1: Newton's Third Law and Rockets

- *drinking straw*
- *oblong balloon (long and thin)*
- *butterfly clamp or paper clip*
- *string, fishing line, or sturdy thread, 6 feet (1.8 m) long*
- *tape*
- *scissors*

Try This!

Work in groups to complete the experiment below. Then, use the **Newton's Third Law and Rockets** Science Log to record your findings.

Procedure:

1. Cut the straw in half to form two 3-inch (7.6-cm) pieces.
2. Blow up the balloon. Close the end of the balloon with the butterfly clamp or paper clip. Be sure air does not escape.
3. Pass the thread, string, or fishing line through both pieces of the straw.
4. Ask your partner to hold one end of the string and tie the other end to a doorknob, so that it slants downward toward your partner. You've just created your flight path.
5. Tape the straws to the balloon as shown.
6. Remove the clamp from the end of the balloon and hold the balloon closed with your fingers.
7. Let your balloon go and observe what happens.

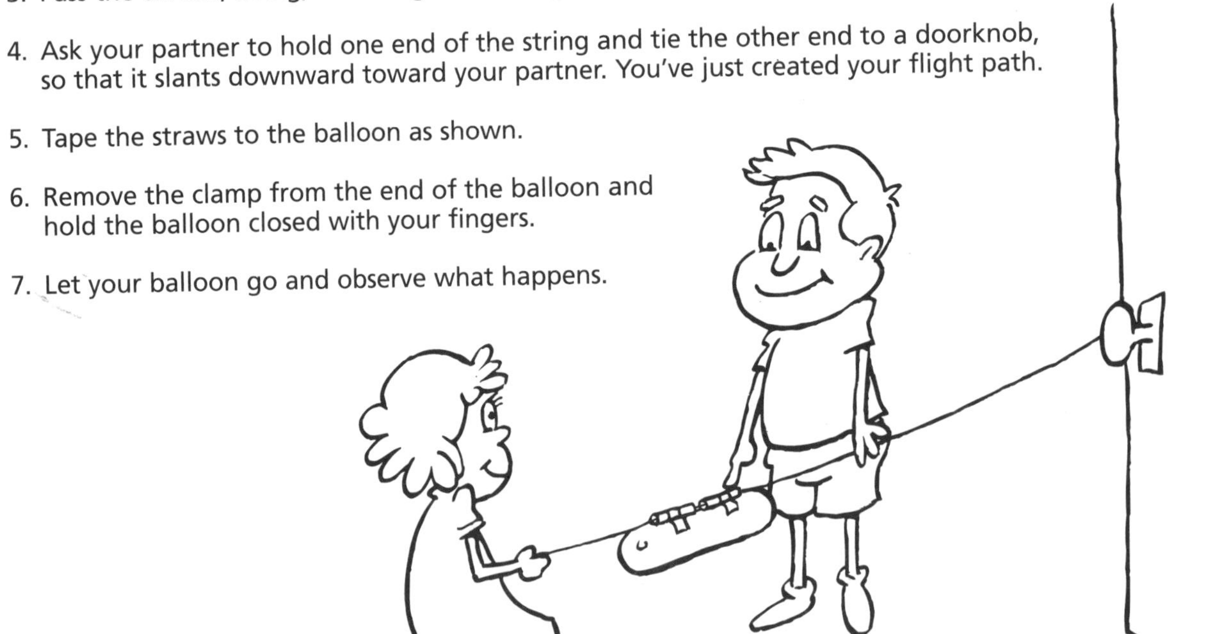

What Happened?

What did your balloon represent? What gave your balloon power? Which direction did the air escape? Which direction did the balloon move? What happened after the balloon ran out of air? How is this like Newton's Third Law?

Name ______________________________

Science Log

Use this sheet to record your observations and work from the **Newton's Third Law and Rockets** experiment.

Question: How does Newton's Third Law explain why rockets lift off? ______________________________

____

____

What I used: ______________________________

____

____

What I did: ______________________________

____

____

____

What happened: ______________________________

____

____

What I learned: ______________________________

____

____

Activity 1: Astronomer Mini-Biography

Create a profile of your favorite astronomer. Fill in the information below. You can select someone you have studied in this lesson or pick someone new.

Astronomer's Name: ______________________________

Birthplace: ______________________________

Date of Birth and Death: ______________________________

____

Famous for the Following Studies: ______________________________

____

____

____

____

____

Interesting Facts Others May Not Know: ______________________________

____

____

____

____

____

____

Lesson 10: Space Exploration

Use this page when you introduce Space Exploration to your students. The fun facts can be used to draw your students into the topic.

They'll Need to Know ...

With a simple rocket blast, we began sending astronauts into space. Since then, we have not been limited to viewing space from Earth. We have moved through space, studying the planets and other celestial bodies much more closely. In fact, astronauts and space probes have landed on some of them!

There were many obstacles that had to be overcome before the first space mission could be a success. A few of these obstacles include overcoming Earth's gravity, dealing with changes in air pressure and gravity, figuring out how astronauts can go about their daily needs in space, and keeping the craft as compact as possible, yet aerodynamic. It's amazing, and we succeeded.

The space program not only launches rockets, but unmanned spacecrafts, including satellites and space probes. A satellite orbits Earth, collecting and sending data to Earth for a variety of uses: weather observation, military information, communications, navigation, and Earth resources data. Space probes collect information about planets, asteroids, comets, and other space objects as they fly by or land on them.

The Space Shuttle and International Space Station are two examples of launches that allow astronauts to stay in space for extended amounts of time to conduct research.

Prove It!

The experiment and activity on the pages that follow provide an experience that explains the basics of rocket propulsion and allows students to create a visual reminder of the incredible gains that have been made in space exploration over the last 40 years. Before you present **Activity 1** on page 72, you will need to explain to students what a timeline is supposed to show and gather images of spacecraft and astronauts in action. Check out the Internet for some good images. You may also want to provide students with a copy of page 69.

Experiment 1: Rockets in Space Teaching Notes: Rocket engines have to be big to launch a spacecraft through Earth's atmosphere. The bigger they are, the heavier they are, and the heavier they are, the more mass holding the spacecraft back. In this experiment, students will notice that their rocket model sped up when it dropped its "back" engine. Explain to students that engineers have designed rockets to drop their burned engines as they shoot into space. Some rockets have three engines in a row, called stages. Each new stage starts when the previous one burns out and drops off. Most of these stages burn up when they tumble back through Earth's atmosphere.

Journal Prompt

Write an essay explaining why you'd like to be an astronaut or why this job would not interest you.

Homework Suggestion

Encourage students to go on a web hunt for information about early space flight, NASA, the Space Shuttle, or the International Space Station. Have them share the addresses for three of the best sites they found with the class.

- The closer you are to the equator when you launch a rocket, the more speed and power it receives. This is because Earth spins with the most speed at 0 degrees latitude. That extra spin provides a little boost of power during lift off.

Lesson 10: Space Exploration (continued)

Here is a brief outline of space missions in the last half century. You may wish to photocopy this information and provide it to students for reference.

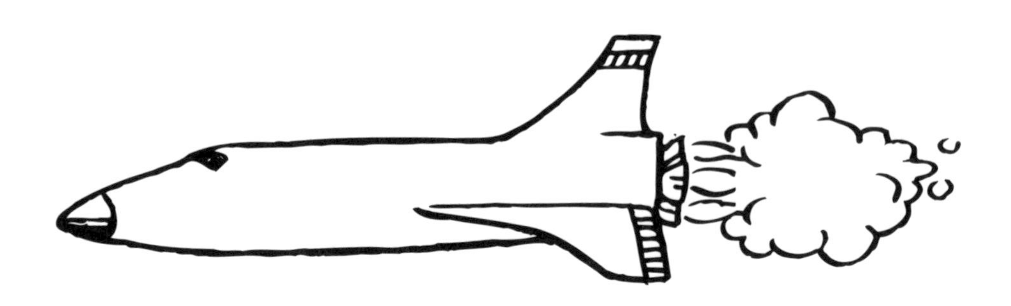

1957: Russia launched *Sputnik*, which becomes the first artificial satellite to be placed in orbit around Earth.

1961: Russian cosmonaut Yuri Gagarin, orbited Earth in *Vostok 1*.

1962: John Glenn became the first American astronaut to orbit Earth, circling the planet three times in *Friendship 7*.

1964: *Mariner* space probe transmitted the first pictures of Mars, marking the first time we took actual photographs of another planet.

1965: Russian cosmonaut Alexei Leonov was the first human to take a space walk outside a spacecraft.

1969: America won the race to the Moon when *Apollo 11* landed there. Neil Armstrong was the first man to actually walk on the Moon from this mission.

1971: Russia created and launched the first space station, *Salyut 1*.

1975: Russian and American spacecrafts docked together in space, allowing cosmonauts and astronauts to spend time together.

1981: The first of NASA's re-useable spacecraft, *Columbia*, took its first voyage.

1984: Bruce McCandless performed a walk in space.

1986: *Challenger* disaster brought NASA's space program to a temporary stop.

1990: The *Magellan* space probe stopped at Venus and spent three years taking pictures of the planet for NASA.

1990: The Hubble Telescope was placed in orbit. The telescope initially had problems, but a repair mission in 1993 got the telescope in working order. The telescope has been sending back incredible images since.

1998: Construction began on the International Space Station (ISS). This project is a joint research venture between the United States and many other countries and will house seven astronauts at a time for extended periods to allow further exploration of the Solar System.

2000: The first successful mission for the ISS was completed. One American astronaut and two Russian cosmonauts stayed on board this craft from October 31, 2000 until March 18, 2001. Two additional missions will take place shortly afterward.

Experiment 1: Rockets in Space

- *2 balloons*
- *empty juice can*
- *measuring tape*
- *2 drinking straws*
- *string*

Try This!

Work with a partner to complete the experiment below. Record your findings on the **Rockets in Space** Science Log.

Procedure:

1. Set up a string between two solid objects, like a chair and a door, and thread two straws through it.
2. Cut the top 4 inches (10.2 cm) off a cardboard juice can. This will make a short tube.
3. Blow up the first balloon through the tube.
4. Hold the neck of the balloon against the inside of the can with one hand, and insert an empty second balloon into the back end of the tube.
5. Blow up the second balloon so it wedges itself in the back end of the tube and prevents the air from escaping out of the first balloon. Pinch the neck of the second balloon to keep the air from escaping.
6. Hold the balloons against the two straws. Have your partner tape one straw to the side of the front balloon, and the second straw to the second balloon.
7. Slide the balloons to the end of the string so your "rocket" can move forward.
8. Release the balloons.

What Happened?

Which balloon emptied first, the front balloon or the back balloon? What happened halfway through the "flight?" How did the rocket's speed change?

Name ______________________________

Science Log

Use this sheet to record your observations from the **Rockets in Space** experiment.

Question: What helps a rocket gain speed? ______________________________

____

____

What I did: ______________________________

____

____

What happened: ______________________________

____

____

Why it happened: ______________________________

____

____

What this illustrated: ______________________________

____

____

____

____

Activity 1: Space Exploration Timeline

Creating a timeline that notes significant accomplishments in space travel offers a wonderful way to see how far we've come and imagine what explorations have yet to be made.

- *scissors*
- *glue*
- *roll of paper*
- *paint, markers, or crayons*
- *research materials*

Procedure:

1. A timeline shows progress and achievements over time.
2. Move into small groups, and wait for your supplies to be distributed.
3. Decide how you would like to present your space exploration information. For example, which decade or decades would you like to show? Also consider adding photos or scale models.
4. End your timeline with questions and ideas for future missions.
5. Discuss your questions about future space travel with your classmates.

Math and the Solar System

There's no better way to enhance learning and make it relevant to students than to tie it with all areas of the curriculum. In this step, you'll find a few fun curriculum-extending activities you might want to try! These pages are meant for you to photocopy, cut by activity, and distribute to your students.

This page is filled with ways you can extend the learning to **Math**.

1. What Would You Weigh on Mars?

It's always fun to compare your weight on Earth to what you might weigh on other planets. The book **The Magic Schoolbus — Lost in the Solar System** by Joanna Cole makes this comparison with each planet the bus visits. Read this book and work with your teacher to convert your own Earth weight to weights on other planets.

2. Chart It!

You know how long one Earth day and one Earth year measure. How about other planets' days and years? Do research to find out and present your data in a chart.

3. How Far from the Sun?

Work with your teacher to find out the distance of each planet from the Sun. Use these measurements and the measurement of Earth's diameter to calculate how many trips around the world it would take to cover the distance between each planet and the Sun. Your teacher can help you.

4. How Large?

Create a graph showing each planet's size. Take this a step further by comparing the size of one planet to another planet. For example, how many Plutos might fit inside Earth?

Planet	Planet's Size	Fun Planet Ratios
Mercury		
Venus		
Earth		
Mars		
Jupiter		
Saturn		
Uranus		
Neptune		
Pluto		

Social Studies and the Solar System

This page is filled with ways you can extend the learning to **Social Studies.** Photocopy this page, cut by activity, and distribute to students.

1. Create a "Sounds From Earth" Tape
The *Voyager* space probes have been traveling through space and sending back information for many years. *Voyager 1* and *Voyager 2* have both left the Solar System and are now traveling through unknown space. Scientists attached a record with two hours of sounds, pictures, and messages from Earth in case other life forms find the probes. Record your own "Sounds from Earth" tape for a probe leaving today. Include important information about Earth and your community on this tape.

2. Discover Asaph Hall
Mars' two Moons were both discovered in 1877 by the American astronomer Asaph Hall. Find out more about his discoveries, the telescope he used, and the time that he lived. Then, write an informational brochure about him for your classmates.

3. My Favorite Planet
Saturn is the ancient Roman god of agriculture who fled to Italy after his dethronement by Zeus as ruler of the universe. Select your favorite planet to research and report on the significance of its name.

4. Creating a Hubble Poster
Hubble Space Telescope's start was problematic. Astronauts were shuttled out to the Telescope while it was orbiting 360 miles (579 km) above Earth. They actually fixed it in space! Find out more about Hubble Space Telescope's history and what it's photographing now by doing research on the Internet. Then, create an informational poster about the Telescope.

5. My Favorite Constellation
Choose your favorite constellation, such as the Big Dipper, Orion, or Leo, and read the myths surrounding that constellation from various cultures. Then, write a myth of your own.

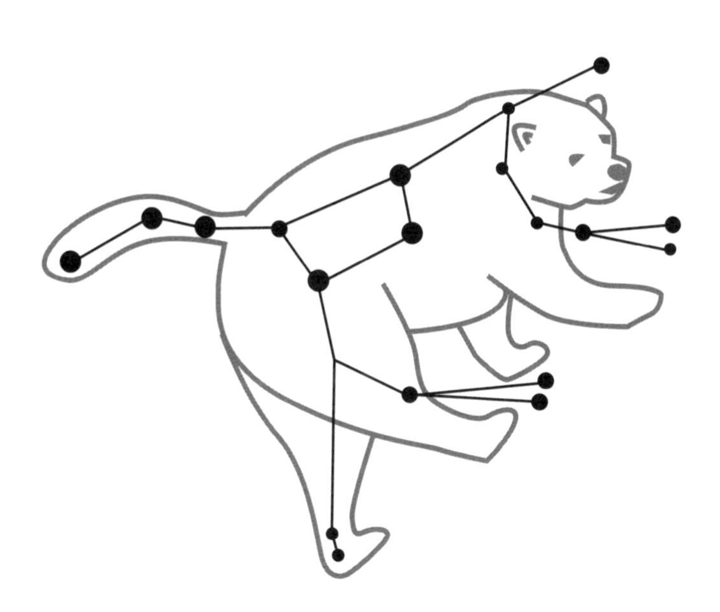

Language Arts and the Solar System

This page is filled with ways you can extend the learning to **Language Arts.** Photocopy this page, cut by activity, and distribute to students.

1. Wordiest Planet Wins!
Here's a game to play with a group of friends. During one minute, players all write down as many adjectives and adverbs they can think of to describe a specific planet. When the minute is up, the writing stops. Then, players take turns reading their lists. If other players have the same word on their lists, they must cross it off. In the end, the winner is the player who has the most words remaining on his or her list.

2. Writing About the Solar System
Write a poem about your favorite planet. You might choose a haiku or an acrostic poem. A haiku uses five syllables for the first line, seven syllables for the second line, and five syllables again for the third line. Acrostic poems spell out a word when the first letter of each line is read from top to bottom. Share your poem with the class. Here is a sample of each:

> Outer Planets (Haiku Poem)
> Saturn with its rings
> Is marvelous in the sky
> As is Jupiter
>
> Mars (Acrostic Poem)
> **M**ore that canyons
> **A** little more than
> **R**ed sand
> **S**imply amazing!

3. Traveling Through Space
Write a travel brochure describing a futuristic vacation to one of the planets. Be sure to describe the terrain, climate, and other aspects of the planet.

4. Writing Science Nonfiction
Write a nonfiction article describing step-by-step how a star is formed. Or, write an instruction sheet for locating a specific constellation. One last idea is to write a list of questions for an inhabitant of our planet and include whether he or she would like to live on Mercury, Venus, Mars, Jupiter, Saturn, Uranus, Neptune, or Pluto.

Reading, Art, and the Solar System

This page is filled with ways you can extend the learning to **Reading and Art.** Photocopy this page, cut by activity, and distribute to students.

1. Taping a Solar System Story
With a group, create a read-along tape for younger children using **The Magic School Bus — Lost in the Solar System** by Joanna Cole. Each student should choose a role to match the characters within the book. Be sure to use an expressive voice as you read. You might even follow up with a list of questions for your audience.

2. Night Sky Creation
Using black bulletin board paper and glow-in-the-dark paint, create a night sky for your classroom complete with constellations!

3. Create a Planetary Creature
Find out what adaptations a life form that looks like a bird would need to survive on Mars. Then, choose a planet that interests you and invent a creature who lives there. Draw the creature and explain which characteristics help it survive on its planet.

4. Water Color Planets
Each of the nine planets has its own distinctive properties and appearance. Create your own visions of the planets with watercolor paints. Hang the paintings as part of an art show with your classmates and then invite your schoolmates to visit the classroom.

TIE IN TECHNOLOGY

Two Great Projects

Technology offers wonderful opportunities for reinforcing learning of all types. In this section, you'll find two great projects that will allow you to take full advantage of all technology has to offer while at the same time strengthen the knowledge gained during the unit of study. Depending on the age group, these activities may be rather advanced. They can, however, be simplified by not using technology or by working through the activities as a whole class. The options are limitless!

1. Create a Multimedia Presentation: Let's Learn About the Planets

A multimedia presentation provides a great way for students to explain the different planets.

Divide students into groups of two or three. Assign each group a planet. Each group should be in charge of preparing slides or cards for their particular planet. Then, all the cards or slides will be compiled into one large presentation.

Next, discuss what information should be included about the planet, such as size, distance from Earth, distance from the Sun, atmosphere, special characteristics, moons, and so on. You should also determine whether students should include bibliography information and if so how to cite their references.

Give groups time to brainstorm their portion of the presentation, and then distribute the **Storyboard** worksheet on page 79. (The groups will more than likely need multiple copies.) If possible, allow the students to spend some time at the computer experimenting with design elements and searching for movies, photos, links or other elements they'd like to include in their presentation. Encourage the use of original artwork and sounds.

Distribute the **Multimedia Presentation Checklist** on page 78. Allow multiple work sessions for planning and the actual creation of the presentation. Then, plan a class "showing" of each group's presentation.

The computer tools you use will depend a lot on what is made available to your school. Some programs that may enhance the project include the word processing and desktopping software available on the market. Other tools include a digital camera, scanner, and even an audiocassette tape. Yet another way to go is to create a poster board per **Storyboard** and use the posters in the presentation. The choices are limitless. However, be sure that students are comfortable using the tools before they start. Also, when students present the project to the class, allow them to use the computer during the presentation to enhance it.

MULTIMEDIA PRESENTATION CHECKLIST

Name ____________________

Planning

- ☐ Have I researched the topic and decided how to show it in a presentation?
- ☐ Have I located outside sources (graphics, sounds, links to web sites, and movies) to use within the presentation?
- ☐ Have I developed a **Storyboard**?
- ☐ Have I determined which tools I need to complete the task?
- ☐ Has each slide or card been designed and numbered?

Content

- ☐ Does my presentation clearly prove a point, explain something, or answer a question?
- ☐ Does the presentation support the content: not too silly if the subject is serious and vice versa?
- ☐ Did I include a table of contents or clear navigation?
- ☐ Are all my references properly cited on a bibliography or reference card?
- ☐ Did I include an "about the author(s)" card?

Design

- ☐ Is it easy to work through the presentation?
- ☐ Is there a good contrast between text color and background color?
- ☐ Are font choices consistent? (Try to use 3 font types or fewer.)
- ☐ Are the sounds, movies, and animations appropriate to the content?
- ☐ Is the text free of spelling, grammar, and punctuation errors?
- ☐ Are the graphics clear?
- ☐ Is the presentation interesting?

Presentation

- ☐ Have I rehearsed the presentation?
- ☐ Have I completed a "dry run" in front of others to make sure the presentation will run smoothly?

MULTIMEDIA PRESENTATION STORYBOARD

Name ______________________________

Use these boxes as you're designing each screen for your presentation on your planet.

1.

2.

3.

4.

5.

6.

2. Create a Web Site: Frequently Asked Planet Questions

This second project will allow you to take full advantage of all technology has to offer while at the same time strengthen the knowledge gained during the unit of study. If your students have already experimented or are ready to learn about web page development, creating a web page is another great way to "show what they know."

The next few pages explain how to create a compelling web site designed to give in-depth information on our Solar System. They do not provide directions on how to build the actual web site. Learning Resources, Inc. offers a wonderful book that explains how to do this. It's called **LER 2282 Technology in the Classroom: Web Page Creation.**

First, spend time viewing web sites. Discuss what makes an effective web site as well as what makes a poor web site. (Use the checklist on page 81 as a guide here.) Introduce the topic of frequently asked questions for your students' web development project, and divide them into groups of two or three.

Next, discuss what you expect as far as content. It might be a good idea to have a class brainstorming session to determine what FAQs your students should include on their web sites. You may decide that students should answer questions about a particular planet or all of the planets together. Either way you go, make sure the web site shows the student groups' link by either the members' first names only or the planet's name. At this time, you should also determine whether students should include bibliography information and if so how to cite their references.

Give students time to brainstorm their web site and then distribute the **Web Site Flow Chart** worksheet on page 82. If possible, allow the students to spend some time at the computer experimenting with design elements and searching for movies, photos, links or other elements they'd like to include as part of their web site. Encourage the use of original artwork and sounds.

Distribute the **Web Design Checklist** on page 81, and allow multiple work sessions for planning and for the actual creation of the web pages. If possible, post the sites to the school server to allow other classes within the school to view the pages. Give students ample time to view each group's site.

WEB DESIGN CHECKLIST

Name ______________________________

- ☐ Is my site's objective clear?
- ☐ Is the subject divided with different subject matter on different pages?
- ☐ Is the text easy to read?
- ☐ Do all links work correctly?
- ☐ Have spelling and punctuation been checked on each page?
- ☐ Is navigation simple to use?
- ☐ Are there links at the bottom of each page so the user can navigate back to the top of the page, the home page, the table of contents, or related information on the subject?
- ☐ Is there a balance between graphics and text?
- ☐ Are font and point size consistent?
- ☐ Is the design consistent?
- ☐ Do all links work correctly?

WEB SITE FLOW CHART

Name ______________________

Use this flow chart to help you think through the design and structure of your web site. Provide notes on buttons, links, design elements, and content.

Page 1

Page 2

Home Page-Title Page

Page 3

Page 5

Page 4

Introduction on Assessment

You've done your job. The content was incredible, the "hands-on" learning opportunities were abundant, and the delivery was no doubt sublime! Now let's see how much actual "learning" took place. There are a number of great ways to assess student learning. We've included some of these methods within the next few pages, complete with rubrics and actual assessments you can photocopy and have students take.

Tests
A well-written exam is the granddaddy of all assessment tools. If you've included everything you want the students to know, a test can be a very reliable measure. We've included two types of tests for this unit: 1.) a Q&A test, and 2.) a multiple choice, matching, and true/false test.

Rubrics
Rubrics allow students and teachers to record their perceptions and opinions. Whenever using rubrics, it's important to encourage honest reporting on the students' part. We've included two rubrics in this section — one for the student and one for the teacher.

Journals
Requiring students to keep a journal as you study a topic serves two purposes. First, it causes the student to recall the information they've just studied. Second, it helps you determine just how much information they took away from the lesson and identify concepts that need further discussion.

The sample journal page included in this book outlines the following areas:

1. *What we studied today.* This encourages students to recap the day's learning.
2. *My experiences with this topic.* Students use this space to share their own experiences with the topic, such as their own studies regarding Earth or space, their thoughts on the Solar System, or the fact they have done the same activity before in another class. If students discuss the latter in this section, encourage them to write about what the activity demonstrates.
3. *Questions I still have.* This is an excellent area for you to identify what students do not understand or to take the learning to the next level. This space allows students to ask any questions they still have surrounding the subject.

Science Logs
Reading through a student's Science Logs will give you clear feedback on whether he or she understood the scientific concept associated with the experiment. Throughout the lessons in **Step 6,** we've included Science Logs for students to fill out when they conduct an experiment. Even though you might provide students with directions for completing each experiment, it's important for them to write down exactly what they did, what materials they used, what the results were, and what they feel the reasons were for the outcome. If what they write is correct and scientifically true, great! If not, you'll know what to review in your upcoming lessons.

A Note About Assessing Projects
While the project in **Step 5: Plan a Project** provides a great way to reinforce learning, they can be tricky to assess. Always monitor each group's performance. Make sure each person is doing a fair amount of the work. If possible, include a peer assessment as part of the overall grade. Be aware that projects don't always cover a complete topic, but rather portions of a topic. Therefore, never base a student's grade for the unit of study solely on a project. We have included some sample project assessment pages throughout this chapter on pages 85-86 for **Step Five: Plan a Project.**

SOLAR SYSTEM JOURNAL

Name ______________________ Date ______________

What we studied today:

____

____

____

____

____

____

____

My experiences with this topic:

____

____

____

____

____

____

____

Questions I still have:

____

____

____

____

____

____

____

Student-to-Student Assessment

Expectations	Actual Performance				
	Never	Sometimes	Frequently	Always	Points
My teammate was helpful.	1 point	2 points	3 points	4 points	
My teammate listened to the ideas presented and participated in group decisions.	1 point	2 points	3 points	4 points	
My teammate contributed a fair amount of work toward the final outcome.	1 point	2 points	3 points	4 points	
My teammate accepted criticism and redirection in a positive manner.	1 point	2 points	3 points	4 points	
Other	1 point	2 points	3 points	4 points	
				Total Points	

Evaluator's Name: ____________________

Comments: ____________________

____

____

Subject's Name: ____________________

Comments: ____________________

____

____

Teacher's Comments: ____________________

____

____

Teacher Assessment

Expectations	Actual Performance & Point Assignment				
	Poor	Okay	Good	Great	Points
Organization	1 point	2 points	3 points	4 points	
Content	1 point	2 points	3 points	4 points	
Mechanics	1 point	2 points	3 points	4 points	
Design	1 point	2 points	3 points	4 points	
Presentation	1 point	2 points	3 points	4 points	
Other	1 point	2 points	3 points	4 points	
				Total Points	

Group Members: ______________________________

Subject's Name: ______________________________

Teacher's Comments: ______________________________

____

Organization: ______________________________

____

Content: ______________________________

____

Mechanics: ______________________________

____

Design: ______________________________

____

Presentation: ______________________________

____

Understanding the Solar System Assessment Test

Name________________________________ Date ________________

True or False

Read each sentence below. Write a T on the line if it is true or an F on the line if it is false.

1. A planet revolves around the Sun. ________
2. A day is the amount of time it takes a planet to orbit the Sun. ________
3. An orbit is the path a planet follows around the Sun. ________
4. Craters are indentations on the Moon's surface. ________
5. The Sun is the largest star in our Solar System. ________

Fill in the Blank

Fill in the blank with the correct word that finishes the sentence.

6. ____________________ is the closest planet to the Sun.
7. ____________________ is called the "Red Planet."
8. ____________________ is the largest planet.
9. ____________________ is the third planet from the Sun.

Multiple Choice

Circle the correct answer that finishes the sentence.

10. Hypatia was:
    A. once a city in Rome B. the first woman astronomer
    C. the astronomer who identified the Sun as the center of our Solar System

11. The first astronaut to orbit the Earth in a spacecraft was:
    A. Neil Armstrong B. John Glenn C. Yuri Gargarin

12. This scientist was famous for his studies of gravity:
    A. Newton B. Galileo C. Aristotle

13. Asteroids are often called:
    A. meteors B. minor planets C. stars

14. This separates the inner and outer planets:
    A. comet B. Asteroid Belt C. Moon

15. This planet is often called the "Sideways Planet:"
    A. Earth B. Saturn C. Uranus

16. This occurs when the Moon passes between the Sun and Earth:
    A. solar eclipse B. snow C. meteor shower

Understanding the Solar System Q & A Assessment

Name________________________ Date ____________

1. In the space provided below, draw and label three planets in the Solar System.

2. Explain how day and night occur. ____________________

3. Explain how a year is determined. ____________________

4. List 2 ways in which Earth is different from other planets.

   a.) ____________________

   b.) ____________________

5. Name 2 astronauts and what they are famous for:

   a.) ____________________

   b.) ____________________

6. List 3 constellations:

   a.) ____________________

   b.) ____________________

   c.) ____________________

Solar System Game Show

It's been an interesting few weeks. You've worked hard to insure student learning. You've required a lot of your students. Everyone, including you, knows a lot more today than you did a few weeks ago. It's time to celebrate your success! What better way to wrap up the unit than with a fun, fast-paced, informative game show?

1. **You'll Need Questions and Answers**
   Assign each student a Solar System topic. If you've taught each of the lessons in **Step 6** of this book, it'll be a good idea to stick with the topics: What is the Solar System?; The Sun; Tilted, Rotating & Revolving; Mercury, Venus, Earth & Mars; Jupiter & Saturn; Uranus, Neptune & Pluto; The Moon; Our Night Sky; History of Astronomy; and The Space Program.

   Photocopy the **Question Form** on page 90. Cut each page to separate the forms, and give each student two forms. Instruct students to write their topic at the top of the form. Then, ask them to write two questions and answers that are related to their assigned topic.

2. **You'll Need a Set and Buzzers**
   Your classroom bulletin board will do fine as the backdrop. Make heading signs for each topic assigned and staple them across the top of the bulletin board.

   Decide on a "buzzer." You could use small bells, whistles, blocks, chimes … ask the school music teacher for ideas. Remember, you'll need four or five — one for each contestant.

3. **You'll Need Contestants**
   Divide the class into groups of four or five. Have each group pick a name for itself having to do with the Solar System. Familiarize students with the topics and encourage them to study hard for the big day!

4. **You'll Need an Audience**
   You know all game shows have audiences to clap and cheer for the contestants. Why should your show be any different? We've included an invitation in this section for you to send out. Fill it out, make multiple copies, and encourage the students to decorate the invitations. Then, distribute the invitations to parents, other classes, school administrators, and friends.

5. **Show Time!**
   The set is ready, the contestants are prepped, and the audience has arrived. It's time to play! Assign one non-contestant student to keep score. Remind the audience to remain silent when a contestant is answering, and remind players of the rules.

   For simplicity, point values will remain constant: 50 points per round. Contestants choose the topic they want when taking their turn. A form will be pulled at random and the question will be read. Contestants must "buzz" in first in order to answer. If the first contestant to buzz in is correct, his or her team gets to choose the next topic. Each team's members take turns playing and go to the end of the line once they have had a chance to provide information. When all the forms have been read, point values are added up, and the team with the most points wins.

Space Game Show Question Form

Topic: ______________________________

____

Answer: ______________________________

____

Question: ______________________________

____

Space Game Show Question Form

Topic: ______________________________

____

Answer: ______________________________

____

Question: ______________________________

____

Space Game Show Question Form

Topic: ______________________________

____

Answer: ______________________________

____

Question: ______________________________

____

You're Invited

Solar System Game Show

Our class has just completed an incredible unit on the Solar System, and now we'd like to challenge each other in a game! Come watch the fun!

What? A Live Solar System Game Show!

When?

Where?

What Time?

Choose from one of the private sessions listed below:

ANSWER KEY

Page 11: Solar System Vocabulary Practice

1. axis
2. Pluto
3. Mars
4. Saturn
5. Galileo
6. constellation
7. Sun
8. Asteroids
9. outer
10. inner
11. craters

Page 12: Solar System Crossword Puzzle

Across

1. revolve
2. asteroids
3. planets
4. Galileo
5. planetarium

Down

6. Mars
7. rotate
8. Neptune
9. Saturn
10. Jupiter

Page 29: Show What You Know: The Solar System

1. Sun
2. Mercury
3. Venus
4. Earth
5. Mars
6. Asteroid Belt
7. Jupiter
8. Saturn
9. Uranus
10. Neptune
11. Pluto

Page 34: Show What You Know: The Sun

1. Sun, star
2. largest, hottest, and brightest
3. mass
4. Gravity
5. hydrogen
6. helium
7. life
8. Sunspots
9. eyes

ANSWER KEY

Page 39: Show What You Know: Tilted, Rotating & Revolving

1. F	6. T
2. T	7. T
3. T	8. T
4. T	9. T
5. F	10. F

Page 57: Show What You Know: The Moon

Across
1. Gravity
2. phases
3. lunar
4. Armstrong
5. Meteoroids
6. Sun

Down

7. days
8. solar
9. craters

Page 62: Show What You Know: Our Sky Night

1. asteroids – rocky bodies smaller than planets that orbit the Sun, often called minor planets

2. astronomer – person who studies the sky and its objects

3. Big Dipper – highly visible part of a larger constellation, often used to locate the North Star

4. black hole – space created when a star dies

5. comet – large, lightweight body that orbits the Sun, contains a head and a tail

6. constellation– grouping of stars that resembles a recognizable image or shape

7. galaxy – large grouping of stars

8. luminous – able to give off light, such as stars do

9. star – hot, glowing ball of gas that gives off light

Page 87: Understanding the Solar System Assessment

1. T	10. B. the first woman astronomer
2. F	11. C. Yuri Gargarin
3. T	12. A. Newton
4. T	13. B. minor planets
5. T	14. B. Asteroid Belt
6. Mercury	15. C. Uranus
7. Mars	16. A. solar eclipse
8. Jupiter	
9. Earth	

Page 88: Understanding the Solar System Q & A

1. answers may vary
2. one day is equal to one complete rotation of the Earth on its axis, during which a point on Earth receives Sun. Night occurs when a point of Earth is turned away from the Sun.
3. A year is equal to one complete orbit of Earth around the Sun.
4. The Earth sustains life and has water.
5. answers may vary
6. answers may vary